# Geology of the Plymouth and south-east Cornwall area

This memoir describes the geology of the 1:50 000 Series Sheet 348 Plymouth.

An introductory section (Chapter 1) describes briefly the relationship of the scenery to the geology; and gives an outline of the geological history.

The Applied Geology (Chapter 2) addresses the major issues of strategic importance to land-use planning and to those people involved with the design and implementation of infrastructural, industrial and housing developments. Key issues dealt with include mineral resources, water resources, aquifer vulnerability and pollution, derelict ground, undermining and made ground; a section on foundation conditions looks at bedrock weathering and slope stability, coastal slope failure, solution effects in the Plymouth Limestone Formation, and superficial deposits, all from a perspective relevant to the civil engineer. This chapter is in itself substantive, illustrating the critical relevance of the new survey information and geological databases to these issues, but it also highlights the extent of the Plymouth data available from BGS to the applied geology professional.

Subsequent chapters describe and interpret scientifically all aspects of the geology of Plymouth and east Cornwall. The nature of the basement and unseen geological foundations of the area are considered in Chapter 3 from the data provided by gravity, magnetic, seismic and magneto-telluric remote sensing surveys in and around the district. A pre-Variscan basement at depths between 8 km and 12 km is suggested with metamorphic rock wedges modelled at shallower depths.

Rocks of Devonian to Permian age and superficial drift deposits are described in chapters 4 to 7, broadly in stratigraphical order. The stratigraphy of the Devonian rocks reveals that they do not form a straightforward sequence, but constitute separate successions that characterise three basins and one intervening high that were variously interconnected at different times. The rock formations record the Early Devonian transition from terrestrial to marine conditions and subsequent deposition in marine environments where lithostratigraphical divisions show complex interrelationships. For example, the Saltash Formation of grey slaty mudstone constitutes parts of the successions of the Looe Basin and of the South Devon Basin, and ranges in age from the late Early Devonian to at least the latest part of the Late Devonian. The Plymouth Limestone Formation and interdigitated volcanic rocks that form the intervening Plymouth High succession are Mid to Late Devonian in age.

Carboniferous rocks of the district (Chapter 5) constitute thrust sheets derived from the Culm Basin to the north. They are predominantly flysch sandstone sequences; the proximal St Mellion Formation, more distal Brendon Formation and the subordinate Newton Chert Formation span the Early Carboniferous (Dinantian), representing sedimentation in an active basin akin to those developed in the Devonian. However, the coarser clastic rocks reflect a changing tectonic regime, indicating uplift of sources to the south that preceded tectonic deformation in the Plymouth district.

The Permian rocks (Chapter 6) are immediately post-tectonic, closely associated with major faults, and rest unconformably on deformed Devonian rocks. They comprise a basal conglomerate and predominant rhyolitic extrusive rocks that are the local surface expressions of igneous activity associated with the Bodmin Moor Granite intrusion.

Quaternary deposits (Chapter 7) are mainly alluvial deposits in the lower river valleys, the hillwash head of the upper reaches of the valleys, and the thick Pleistocene head on the lower raised marine platforms along the coast.

Devonian and Lower Carboniferous igneous rocks of the district are part of a province-wide expression of magmatic activity (Chapter 8). They are dominantly basic extrusive and intrusive rocks, with minor occurrences of acidic tuffs. The igneous rocks are largely within-plate alkaline basalts derived from the mantle, and their emplacement was closely linked with rifting and basin generation. In contrast, the Permian igneous rocks comprise essentially rhyolitic lava and associated quartz-porphyry dykes, with the lava being the only unequivocally in situ acid volcanic rock in the Permian of the province.

Mineralisation of three main types is recognised in the area (Chapter 9): pre-granite syngenetic and epigenetic concentrations, granite related hydrothermal veins and post-granite crosscourses. The first type, linked with the various phases of basaltic volcanism, was subject to remobilisation, concentration and re-emplacement during deformation and metamorphism. Further concentration with new metal input from the granite-related hydrothermal systems gave rise to the main lodes of the area. The historically productive crosscourse deposits of the Plymouth district are attributed to basinal saline fluids that circulated in the Triassic Period.

The structures developed during the main phases of Variscan compressional and extensional deformation and the regional metamorphism of the rocks are described in Chapter 10. Detail of chronology, style, attitude and distribution of folds, cleavages, thrusts and other faults permits correlation across domains of opposed facing, differing metamorphism and deformation timing.

The geology revealed by this survey and linked research programme is synthesised in Chapter 11. All aspects of the solid rock geology are interdependent and are drawn together in the context of the evolving Variscan passive margin, and its subsequent history.

Sources of data relevant to the district and adjacent areas is listed in the final part of the memoir under *Information Sources* and there is a comprehensive list of references.

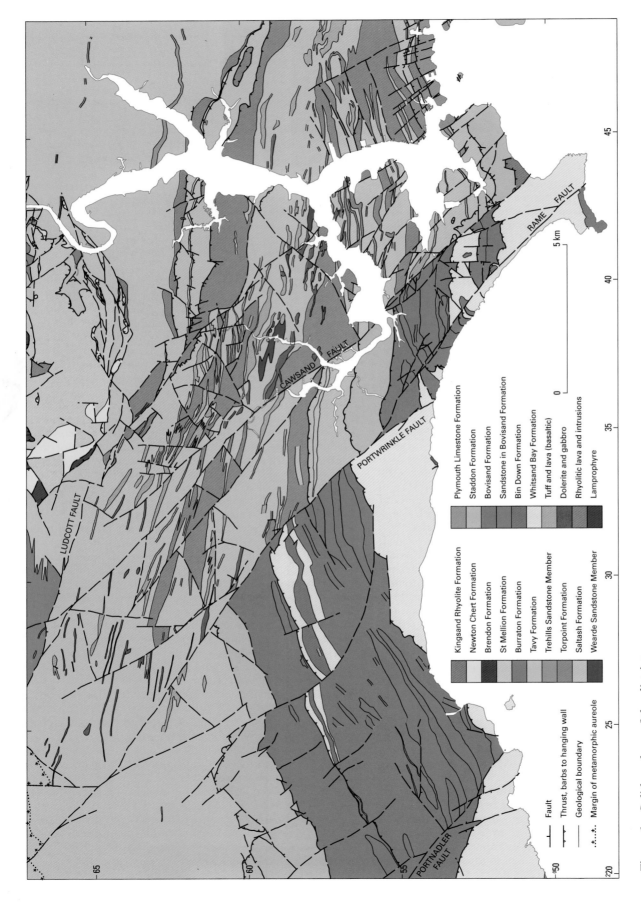

**Figure 1** Solid geology of the district.

Fault

Thrust, barbs to hanging wall

Geological boundary

Margin of metamorphic aureole

Kingsand Rhyolite Formation

Newton Chert Formation

Brendon Formation

St Mellion Formation

Burraton Formation

Tavy Formation

Trehills Sandstone Member

Torpoint Formation

Saltash Formation

Wearde Sandstone Member

Plymouth Limestone Formation

Staddon Formation

Bovisand Formation

Sandstone in Bovisand Formation

Bin Down Formation

Whitsand Bay Formation

Tuff and lava (basaltic)

Dolerite and gabbro

Rhyolitic lava and intrusions

Lamprophyre

LUDCOTT FAULT

CAWSAND FAULT

PORTWRINKLE FAULT

PORTNADLER FAULT

RAME FAULT

5 km

0

BRITISH GEOLOGICAL SURVEY

B E LEVERIDGE
M T HOLDER
A J J GOODE
R C SCRIVENER
N S JONES and
R J MERRIMAN

CONTRIBUTORS

*Palaeontology*
M T Dean
A McNestry
I P Wilkinson
S G Molyneux

*Sedimentology*
J R Davies

*Geophysics*
I F Smith

*Engineering geology*
S M Fenwick
T P Gostelow

*Hydrogeology*
M A Lewis

# Geology of the Plymouth and south-east Cornwall area

Memoir for 1:50 000 geological Sheet 348
(England and Wales)

London: The Stationery Office    2002

ISBN 0 11 884560 8

*Bibliographical reference*

LEVERIDGE, B E, HOLDER, M T, GOODE, A J J, SCRIVENER, R C, JONES, N S, and MERRIMAN, R J. 2002.   Geology of the Plymouth and south-east Cornwall area.   *Memoir of the British Geological Survey,* Sheet 348 (England and Wales).

*Authors*

B E Leveridge, BSc, PhD, CGeol, FGS
M T Holder, BSc, LLM(env.law), PhD, CGeol, FGS
A J J Goode, BSc
R C Scrivener, BSc, PhD, CGeol, FGS
*British Geological Survey, Exeter*

N S Jones, BSc, PhD
R J Merriman, BSc, CGeol, FGS
*British Geological Survey, Keyworth*

*Contributors*

M Dean, BSc, MPhil
*British Geological Survey, Edinburgh*

S M Fenwick, BSc, MSc
T P Gostelow, BSc, MSc, PhD
A McNestry, BSc, PhD
S G Molyneux, BSc. PhD
I F Smith, BSc, MSc
I P Wilkinson, BSc, MSc, PhD, CGeol, FGS
*British Geological Survey, Keyworth*

J R Davies, BSc, PhD, CGeol, FGS
*British Geological Survey, Aberystwyth*

M A Lewis, BA, MSc
*British Geological Survey, Wallingford*

Printed in the UK for The Stationery Office
TJ 85011   C6   3/02

*Front cover*   View over the Hamoaze and River Tamar valley, looking north-north-west from the Staddon Formation scarp, Mount Edgecumbe Country Park [450 520]. Devonport Dockyard is to the centre right, Saltash and the Tamar road and Brunel railway bridges are in the centre distance, and the St Mellion outlier forms the high ground in the distance (GS 508). (*Photographer*   A J J Goode).

# CONTENTS

## FIGURES

TABLES

PLATES

# PREFACE

The geological map Sheet 348 Plymouth includes much of the city of Plymouth, together with part of south Devon and a large part of the Caradon District of east Cornwall. The valley of the River Tamar (the county boundary) is designated an *Area of Outstanding Natural Beauty*. The sheltered deep-water anchorages of The Sound and the Hamoaze, at the estuary of the Tamar, have given Plymouth an historically important role as one of the country's major naval ports and support establishments. However, recent changes have seen a reduction of defence forces and the decline of the naval support industry in Plymouth, hitherto so important to the sustenance of the city.

The resurvey of the Plymouth district was undertaken to provide modern geological information, and a comprehensive understanding of the geology. Together, the new information and concepts will provide a sound basis for future land use and development planning, as well as urban and regional regeneration.

In the past a wide range of minerals has been exploited within the district. This activity has now ceased, leaving a legacy of contaminated land and undermining. Such factors must be considered in any future development. Other issues discussed in this memoir of interest to civil engineers and planners are foundation conditions such as bedrock weathering and slope stability, coastal slope failure, solution effects within the limestone; water resources, aquifer vulnerability and pollution are also discussed.

The original detailed survey of Plymouth and adjacent one-inch sheets, just over a century ago, was by one of the Survey's great field geologists, W A E Ussher, whose observation and records have withstood the test of time. In order to make significant advances in understanding the geology of this complex area, this re-survey has had to be considerably more detailed, and well supported by multidisciplinary studies. The results are embodied in the new 1:50 000 Series map, in the memoir and in the supporting data that forms part of the BGS national geoscience archive. The great detail of geological information now available on stratigraphy and rock structures combined with a wealth of borehole data and, most importantly, the expertise developed in BGS, means that a sound, informed and independent geoscientific input into all infrastructural developments and environmentally linked programmes within the district is now possible. Indeed, significant contributions have already been made during the latter stages of the survey to major pipeline and tunnelling operations in the area and to projects linked to the future development of Devonport Dockyard.

This survey has demonstrated for the first time the existence of a Palaeozoic passive margin and charted its evolution within the Variscan orogenic cycle. Mineralisation, fundamentally linked to igneous activity, can now be placed and explained in the appropriate tectonic setting of continental margin rifting and post-collision granite intrusion, with remobilisation and emplacement linked to tectonic reworking and post-tectonic fluid circulation.

This memoir provides not only a scientific explanation of all the interdependent aspects of the Variscan geology of the Plymouth area but also the key to a wealth of detailed and practical information relevant to future development of the district.

David A Falvey, PhD
*Director*

*British Geological Survey*
*Kingsley Dunham Centre*
*Keyworth*
*Nottingham*
*NG12 5GG*

## Acknowledgments and survey details

The memoir has been written largely by B E Leveridge, M T Holder and A J J Goode, with R C Scrivener (Mineralisation), N Jones (Sedimentology) and R J Merriman (Metamorphism, Petrography and Geochemistry) making substantive contributions of script and to the understanding of the geology of the area. The manuscript was edited by P Stone, R D Lake and A A Jackson. The figures were produced by R J Demaine, P Lappage and G Tuggey, BGS Cartography, Keyworth.

The mapping and related research on the Plymouth Sheet has been a multidisciplinary effort and those contributing specialist scripts and reports heavily drawn upon in the memoir are M Dean, A McNestry, I P Wilkinson and S G Molyneux (Biostratigraphy), J R Davies (Limestone sedimentology), I F Smith (Geophysics), and M Lewis (Hydrogeology).

In addition to the named contributors, BGS staff making valued inputs during the course of the mapping programme include B Owens and N Turner (Palaeontology), R W O'B Knox and B Humphreys (Sedimentology), D Beamish and J W F Edwards (Geophysics), S J Kemp and S Prior (Metamorphism), and C P Smith.

The expert contributions made by the following external specialists are also very gratefully acknowledged.

Pierre Bultynck, Royal Belgian Institute of Natural Sciences, Brussels, kindly commented upon and confirmed conodont identifications by Dean (BGS) recorded in several biostratigraphy reports listed in the Information Sources section.

A Dean, University of Exeter, provided the results of his analysis of the palynomorph and acritarch assemblages from samples collected during the survey.

L N Warr, University of Heidelburg, made illite crystallinity determinations on many of the slate and cleaved mudstone samples, required to produce the metamorphic map figured in Chapter 10.

S A Smith, (formerly BGS), Shell Exploration, Netherlands, provided key insights into the sedimentology of Lower Devonian rocks at the commencement of the mapping programme.

C J Burton, University of Glasgow, generously supplied mapping data and information on his palaeontological determinations in the Liskeard area, for use during the survey.

R G Thomas is thanked for supplying detailed information over several years on temporary excavations and exposures in Plymouth and the neighbouring areas.

B V Cooper collected minerals samples from Tordown and Tredinnick mines.

Grateful thanks are also due to the many landowners, farmers, householders, civil and military authorities, for property access and helpful assistance during the survey.

The map that the memoir describes was surveyed in large part by M T Holder, B E Leveridge and A J J Goode. C M Barton and B J Williams made contributions to the mapping during the course of the programme.

## Notes

Throughout this memoir the word 'district' refers to the area covered by the 1:50 000 Series Sheet 348 Plymouth.

Figures in square brackets are National Grid references; the Plymouth district lies within the 100 kilometre square SX.

Numbers preceded by the letter A refer to BGS photograph collection.

The bracketed 1:10 000 scale Ordnance Survey sheet reference with succeeding oblique and number is a British Geological Survey archive reference.

Authorship of fossil species is given in the *Fossil inventory*.

Numbers preceded by the letter E refer to thin-sections in the collection of the British Geological Survey, Keyworth, and those preceded by SDP are held in the Exeter office.

For geological enquiries and addresses see *Information Sources*.

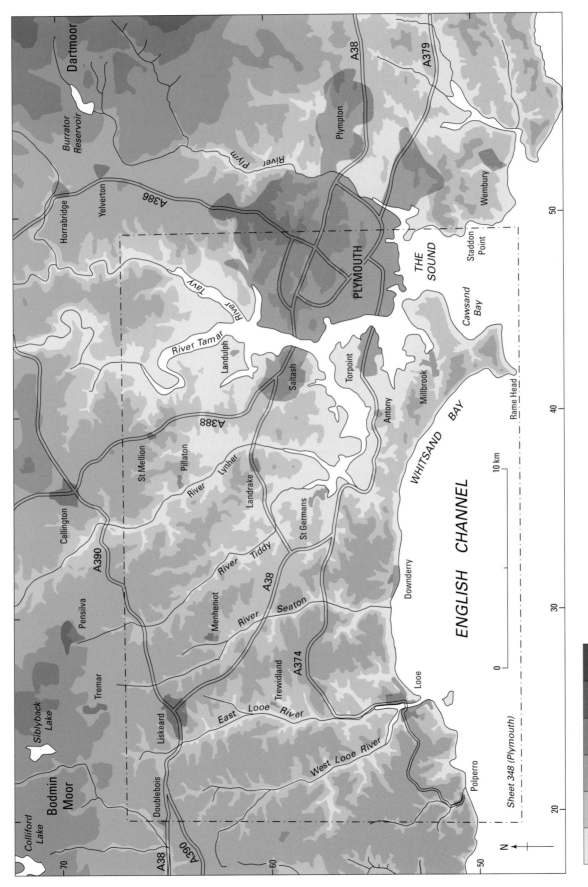

**Figure 2** Topographical map of the district and surrounding area.

# ONE

# Introduction

The district described in this memoir extends from Polperro and Dobwalls, near Liskeard, in the west, to the city of Plymouth in the east, across the River Tamar that forms the county boundary between Devon and Cornwall. In the north, there is elevated ground that lies between the Dartmoor and Bodmin Moor granites (Figures 1, 2). Apart from the fine coastal scenery along the English Channel, it includes designated areas of outstanding natural beauty such as the Tamar valley and the estuary of the St Germans or Lynher River.

The topographical features in the area are closely linked to the strike of bedding and cleavage in the underlying rocks. The more resistant lithologies, such as dolerite, volcanic rock, chert and sandstone, form ridges. Positive features are also a characteristic of the hanging walls of some major thrusts, where lithologically similar sequences are juxtaposed. Such features are offset by cross-cutting strike-slip faults and oblique normal faults. In the western half of the district steep-sided river valleys are cut along the lines of faults over considerable distances. By way of contrast, the ria system of the River Tamar, particularly in its southern reaches, is flanked by low cliffs and the courses of its tributaries are only very locally constrained by faults. The River Tamar rises in north Cornwall, not far from the coast, and flows southwards, in a deeply incised valley system, through the elevated land between Dartmoor and Bodmin Moor.

The coastal cliffs along this part of the English Channel are commonly up to 100 m or more in height, except where there are remnants of a Pleistocene platform, for example west of Penlee Point [443 487] and south of Looe [255 523], or where the cliffs are degraded by landslip, as to the east of Millendreath [285 544].

## OUTLINE OF THE GEOLOGY

This district is one of great geological complexity and poor inland exposure so that research has progressed slowly since the broad stratigraphical subdivisions were established late in the nineteenth century. The recent resurvey which included detailed inland mapping has provided substantial new data to accompany the new map, shown here in a simplified form (Figure 1).

The Devonian and Lower Carboniferous (Dinantian) rocks of this district (Figure 1) were deposited on the northern side of an evolving (Rhenohercynian) ocean basin, which stretched eastwards across northern Europe. In this district, crustal extension and faulting occurred during the Early Devonian, and the exposed Early Devonian sedimentary rocks record a transition

from a continental, subaerial environment to a submarine environment. Continued extension until the end of the Dinantian produced narrow, elongate rift basins that developed and filled sequentially northwards. On the intervening highs, limestone and volcanic rock accumulated from the Mid to Late Devonian times. Because of this complex basin topography, there is no regional succession, but different successions characterise each basin and high (Table 1).

It was the coexistence of shallow and deep water sediments in the Middle and Upper Devonian strata of south Devon that attracted previous workers to make comparisons with the concept of a basin (becken) and rise (schwellen) depositional framework, established for the coeval Rheinisches Schiefergebirge of Germany where a primary horst and graben architecture was later recognised. This led to the recent proposal of a half-graben model for the rocks of north Cornwall and their equivalents across the peninsula.

In this district, faults commonly separate different suites of sedimentary rocks which strongly suggests that the basin margins were fault-controlled. These structures were probably curvi-planar (listric) faults, which caused block rotation, and bounded full or half grabens. Four main basins have been recognised in this district: the Looe Basin and South Devon Basin are composites of half-grabens whereas the Tavy Basin and possibly the Culm Basin show the overall geometry of a full graben. The distribution of the various successions is shown in Figure 3.

In the Looe Basin, the thick Early Devonian sediments show a transition from lacustrine and fluvial to fluviodeltaic and marine conditions in a depositional basin characterised by rapid but phased subsidence. The highest part of the succession preserved shows that the basin was later largely shielded from an input of coarser sediment from the north by the development of the Plymouth High and associated South Devon Basin.

The Plymouth High, to the north, is characterised by the presence of stromatoporoid and coral reefs and records a period of carbonate shoaling in shallow waters that existed in Mid Devonian times. The close association of limestone with volcaniclastic rock suggests that reef growth was founded on volcanic accumulations. These relatively thin sediments occur within two juxtaposed major thrust slices, in a structural position such that this high separated the Looe Basin from another basin with mostly younger sediments, to the north.

This northerly South Devon Basin (and its extension westwards to the Trevone Basin of north Cornwall) contains marine, grey, slaty mudstone that tends to dominate this thick sequence; comparable coeval sediments in the Looe Basin demonstrate some degree

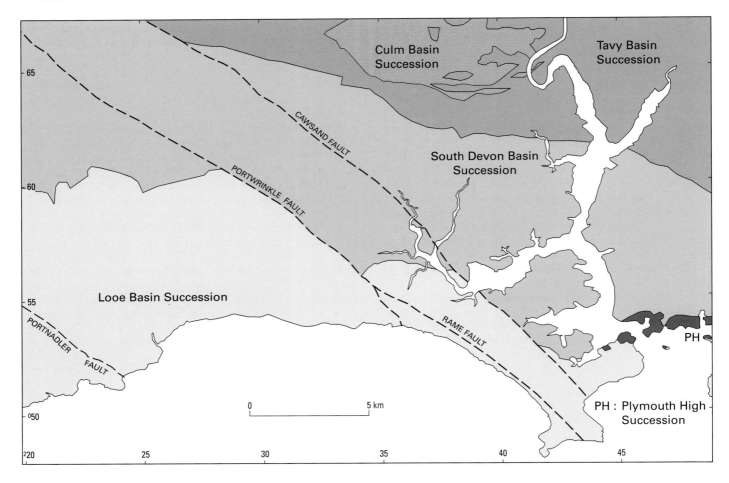

**Figure 3** Distribution of basin and high successions in the Plymouth district.

of overlap of succession and connection between the two basins. The sediments in the South Devon Basin also comprise turbiditic sandstone, more commonly in the lower part of the sequence, and isolated beds of limestone, as well as tuff and basalt lava. In the Landulph area, the highest part of the preserved succession is seen to contain chert of possible early Carboniferous age, and sporadic limestone beds. Interdigitating with these grey lithologies, there is a red and green sequence of mainly fine-grained sediments that has been mapped as a separate Torpoint Formation of Late Devonian age. This formation occupies discrete tracts in the eastern part of the district and shows splitting and thinning, dying out westwards. Detailed studies have shown that these rocks are not altered varieties of the grey rocks but that they had a separate source.

The Landulph High, which separates the South Devon and Tavy basins, is not as clearly defined as the Plymouth High, but limestone turbidites and a slumped limestone mass in the basinal sediments near Landulph indicate a nearby source to the north. The Late Devonian to early Carboniferous sediments of the Tavy Basin comprise dominantly green and grey slaty mudstone with some turbiditic sandstone.

In the early Carboniferous, the first coarse clastic sediments, derived from a rising and tectonically deforming tract to the south, were deposited to the north of the Plymouth district in an evolving marine Culm Basin. In this district, these thick formations occur in entirely thrust-bounded, nappe units and comprise a sequence dominated by turbiditic sandstone and a correlative mainly mudstone division. The structurally discrete Newton Chert Formation, within this fault-complex, contains radiolarian chert that correlates regionally with a phase of silica saturation associated with early Carboniferous volcanicity. Manganiferous stratiform ores within the formation are directly linked with this last phase of Devonian/Carboniferous volcanicity of the region.

The first major regional deformation of these Devonian and Carboniferous rocks is thought to have migrated northwards through south-west England during the Carboniferous. It followed a continental collision which caused the overthrusting (obduction) of the Lizard ophiolite in the Late Devonian/early Carboniferous. This concept of diachronous deformation is based upon the K–Ar metamorphic age determinations of slate made throughout the peninsula, linked to structural studies, and the dynamic stratigraphy of south Cornwall.

In this district, the deformation which occurred between 345 and 325 million years BP in Viséan and earliest Namurian times caused extensive thrusting, folding and a variable intensity of cleavage. The extensional faults which had controlled the form of the basins became thrusts, thus accommodating basin inversion. Generally the largest folds (antiforms) developed adjacent to these thrust faults that delimited major rock suites. The geometry of fold and thrust fault vergence is a function of original basin form and attitude of the basin-controlling faults. Thus the southerly inclined listric faults bounding the Looe and South Devon basins initiated northerly vergent thrust and fold structures, whereas a northerly inclined fault at the southern margin of the Tavy Basin produced southerly vergent structures. Mica crystallinity values which reflect the metamorphic grade generally directly relate to the age of the rocks.

Towards the end of the Carboniferous, a second major compressive deformation, part of a synchronous Europe-wide episode, affected the region, causing the full inversion of the Culm Basin. In the Plymouth district this phase produced secondary folds oblique to earlier structures and out-of-sequence thrusting. Transpressive stress affected the early basin-controlling east–west faults, and their linked later structures, to produce dextral shear fabrics. The major southward verging folds and thrust faults on the southern margin of the former Culm Basin were probably reactivated at this time. The structural edifice built by the overall composite inversion was subject to later gravitational collapse, with thrust sheets extending further southwards into the district, and the formation of southerly vergent folds.

Following the last compressive episode, the region was subject to north–south extension which caused backslip on earlier thrusts and extensive normal faulting. Fault-bounded basins developed against earlier strike-slip faults, as at Kingsand, where coarse clastic detritus was rapidly deposited in early Permian times. The regional release of pressure was accompanied by the generation of granitic magma at the base of the crust and the intrusion of major granites to the north of the district. This intrusion was accompanied by acid volcanism, as at Cawsand Bay in this district, and by tin and tungsten mineralisation. Regional hydrothermal activity also scavenged chemical elements from the country rocks, most importantly copper, which were then emplaced as ore mineral veins, termed main lodes. Some of these are present in the north-west of the district.

There is little geological evidence within the district of events in post-Palaeozoic times. The ore minerals that are present in north–south fractures, as crosscourse veins, are thought to have been introduced by saline fluid circulation, associated with the deposition of thick Triassic deposits, no longer preserved hereabouts. Dextral displacement of the Bodmin Granite aureole, and the rotation of some adjacent fault blocks along earlier major north-west-trending faults of the district are probably attributable to Cainozoic earth movements.

In much more recent times, marine retreat from a putative high sea level in the early Cainozoic (Tertiary) has left a record of still-stand features, a 'staircase' of wave-cut platforms, at varying intervals. Those platforms below 40 m above OD are related to the Quaternary Period. Marine cave systems, developed in Devonian limestone at a variety of levels, relate to stillstands in the overall regression and contain sparse deposits that include animal remains. Late Pleistocene (Devensian) head deposits of periglacial origin, up to 35 m thick, mantle the lowest marine retreat platforms around the coasts. Low sea levels in this cold climatic phase caused the river systems to cut down to levels of about 40 m below OD. With the subsequent amelioration of climate in the Holocene, sea level rose and alluvial deposits now partly fill the buried valleys present in the Tamar estuary and environs.

4

# TWO

# Applied geology

In this chapter the geological factors relevant to land-use planning and development within the district are reviewed.

## KEY ISSUES

Geological factors have played, and continue to play, an important role in the economic development of both Plymouth and Liskeard and the area surrounding them. The exploitation of mineral resources in the district has left a serious, although geographically somewhat restricted, legacy of derelict and contaminated land. The characteristics of the rocks underlying the district and the effect of natural processes on them present problems for current and future development. All of these are issues of high priority for land-use planning and for the design and implementation of infrastructural, industrial and housing developments.

- *mineral resources*: past exploitation, remaining resources
- *water resources*: surface water, groundwater
- *aquifer vulnerability and pollution*: water chemistry, contamination of groundwater, leachate contamination from landfill sites
- *derelict ground*: contamination from disturbed mine dumps, contamination of surface waters from mine tailings, potential contamination from waste tips
- *undermining*: subsidence, groundwater pollution
- *made ground*: extent, characteristics
- *foundation conditions*: weathering zonation, rock strength, slope stability, dissolution of limestone, coastal stability

## MINERAL RESOURCES

A wide range of minerals have been exploited within the district. They include limestone for building stone, agricultural lime, cement and aggregate; dolerite, rhyolite, chert and sandstone for roadstone and building and slaty mudstone for building, roofing slates and brick making. Almost all of the extraction of these bulk minerals has now ceased. Metalliferous minerals have been mined within the district (Figure 4) for lead, silver, zinc, copper, tin, antimony, arsenic, baryte and fluorspar (Table 2). This mining has also now ceased.

Further information on the mineral resources of the district is available from BGS. This may be of help, to enable informed planning decisions regarding the environmental effects of mineral extraction, reclamation of sites and the avoidance of sterilisation of mineral resources.

**Limestone** has been extensively quarried from the Plymouth Limestone Formation for building stone, agricultural lime and cement throughout the southern part of urban Plymouth. In places these quarries have become amalgamated leaving small 'islands' and narrow 'walls' of unquarried limestone, particularly to the east of the district in eastern Plymouth. There are no limestone quarries currently active within the district, but limestone for aggregate and cement is still quarried in eastern Plymouth at Pomphlett quarry [5120 5430] and Moorcroft quarry [526 541].

The **doleritic** and **picritic** intrusive rocks of the district have been quarried for roadstone and aggregate in a number of generally small- to medium-sized quarries, now disused, dispersed around the district. The size of these quarries was largely restricted by the size and shape of the intrusions themselves with the largest of the quarries within the extensive picrite intrusion at Clicker Tor [2860 6135]. Basic igneous rocks are currently being worked only at Lean quarry [2650 6140] where basaltic pillow lavas and dolerite intrusions in the Saltash Formation are extracted for random walling and landscaping stone.

Small quarries within the lower Viséan **chert** have been worked for roadstone and thick **sandstone** beds within the St Mellion Formation, the Trehills Sandstone Member of the Tavy Formation, and the Staddon Formation, were formerly worked as buff-weathering freestone for building. None of these quarries is currently active.

The Devonian **slaty mudstone** rocks were much used locally for building and walling. Most of this stone was extracted from small local quarries which are now disused and many backfilled or developed for industrial or housing use. Only one quarry in these rocks, at Lantoom [2245 6490], is currently active extracting dark grey slaty mudstone from the Saltash Formation for decorative walling, landscaping and paving. Sources of local building stone for restoration or for ornamental use on new buildings are therefore strictly limited. Table 3 summarises the locations and building stones potentially available in some of the larger disused quarries which might still be reworked for this purpose. Permian **rhyolite** was quarried in several small quarries around Kingsand [4327 5096] for roadstone.

Devonian slaty mudstone has also been used for the manufacture of poor quality bricks. In the 19th century there was a small brickyard at Sutton Harbour [4915 5417] working the surface weathered slates at the northern edge of the Plymouth Limestone. At Stoke [4642 5571], the Torpoint Formation was quarried for brick making. The sites of both of these brickyards have long since disappeared beneath the urban development

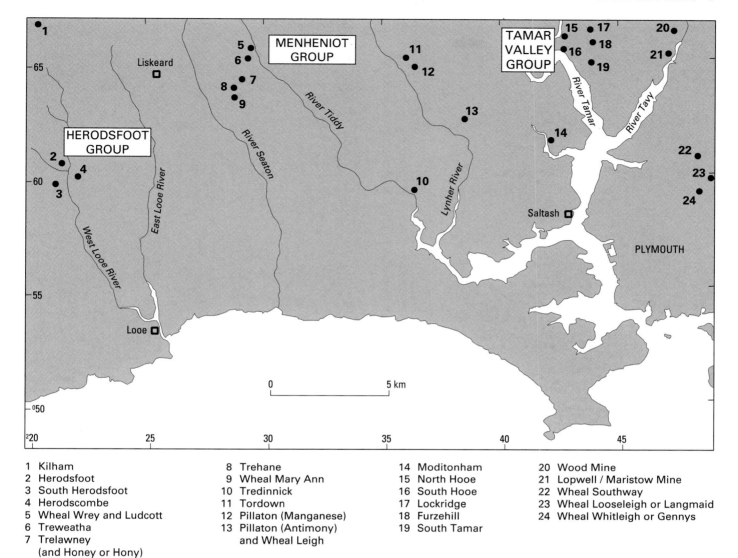

1  Kilham
2  Herodsfoot
3  South Herodsfoot
4  Herodscombe
5  Wheal Wrey and Ludcott
6  Treweatha
7  Trelawney
   (and Honey or Hony)

8  Trehane
9  Wheal Mary Ann
10 Tredinnick
11 Tordown
12 Pillaton (Manganese)
13 Pillaton (Antimony)
   and Wheal Leigh

14 Moditonham
15 North Hooe
16 South Hooe
17 Lockridge
18 Furzehill
19 South Tamar

20 Wood Mine
21 Lopwell / Maristow Mine
22 Wheal Southway
23 Wheal Looseleigh or Langmaid
24 Wheal Whitleigh or Gennys

**Figure 4**  Location of metalliferous mines within the district.

of Plymouth. Much larger brickyards were sited near Inswork [4255 5265 and 4325 5237] on the shore of Millbrook Lake where a combination of grey slaty mudstone and tuff of the Saltash Formation was used. Coastal **head** was also used for local brickmaking at Hannafore Point [2555 5235] near Looe.

## WATER RESOURCES

All of the public supplies for the Plymouth district are provided from surface water sources, although the more rural parts of the district use a combination of surface and small groundwater sources.

Additional information on wells and groundwater, for the assessments of pollution risks from industrial and waste disposal sites, as well as for locating further local water supplies, is available from BGS.

Historically, Plymouth derived its water supply from shallow wells sunk into the limestone and shale as the streams within the city are tidal. However, after Drake's Leat was constructed in 1590 most of these (which still survive as street names e.g. Buck Well and Lady Well), fell into disuse (Roxburgh, 1983), although they were still used in dry summers or cold winters when the leat dried up or froze. The leat brought water nearly 30 km from the Meavy catchment to Plymouth, Stonehouse and Devonport; initially in an open channel and later in pipes (Ussher, 1907). During Victorian times and the early part of this century, several boreholes were constructed into the Plymouth Limestone Formation. However, none of these are now in use. In 1898 the Burrator Reservoir was completed, and Plymouth and the surrounding area began to derive their supplies from a combination of this reservoir, the Devonport Leat and the River Tavy. The Roadford Reservoir has only recently been completed.

**Table 2**  Mines that worked stratigraphically related mineral deposits.

| Mine/Location | Associated mines | Minerals | Notes | Production | Working period |
|---|---|---|---|---|---|
| Pillaton | Wheal Leigh, Leigh Durrant Mine | Stibnite | Associated with Upper Devonian basic volcanic rocks | 132 tons antimony ore (1819–1821)[1] | Commenced 1572 |
| Tredinnick | | Stibnite | Associated with Upper Devonian basic volcanic rocks | 14 tons 70% lead ore (1876)[2] | working in 1830s and 1891 |
| Pillaton | | | Associated with Viséan chert | Manganese[3] | |
| Tordown | | Rhodochrosite and Rhodonite | Small underground workings at Tor Wood associated with Viséan chert | Manganese[3] | |

Mines working east–west-trending steep mineral lodes

| Mine/Location | Associated mines | Minerals | Notes | Production | Working period |
|---|---|---|---|---|---|
| Kilham | Pensilva Tine Mine Wheal Coryton | Pyrite Cassiterite, Arsenopyrite Chalcopyrite Galena Siderite Sphalerite | Lode 50 mm–2 m wide possibly extension of St Neot's Mine[4] | 3.5 tons tin concentrate 1908–1909 | 19th century |

Mines working north–south-trending steep mineral lodes

| Mine/Location | Associated mines | Minerals | Notes | Production | Working period |
|---|---|---|---|---|---|
| North Hooe | North Tamar Mine Wheal Hancock Tamar Consols (with South Hooe Mine) | Galena Sphalerite Fluorite | Lode 0.46 m–1.22 m wide. Vertical to 75°E | 1200 tons lead silver ore (1846)[5] | 1842–1855 1886[6] 1901–1902[6] 1906[4] |
| South Hooe | Tamar Silver-Lead Mine Tamar Consols (with North Hooe Mine) | Galena Fluorspar | | 326 300 oz silver (1845–1876) 780 tons fluorspar (1879–1882) | before 1935–1885 |
| Lockridge | Goldstreet, Whitsam East Tamar Consols (with Furzehill) | Pyrite Sphalerite Fluorite | | 650 oz silver 620 tons fluorspar 20 tons pyrite (1870–1876) | 1866–1876[7] |
| Furzehill | Whitestone (Whitsam?) Down Mine, East Tamar Consols (with Lockridge) | Galena Sphalerite Pyrite, Fluorite[4] | Flooded by South Tamar Mine collapse 1856 | 2580 tons 69% lead ore 19 530 oz silver (1845–1856) 1400 tons fluorspar (1857–1858)[4] | working in 1845 until at least 1858 |
| South Tamar | Cleave, Birch, South Tamar Consols | | Lode strike 012°N and dips 85°E Flooded in collapse in 1856 | 7140 tons 64% lead ore 350 tons fluorspar 262 470 oz silver (1849–1856) | working before 1817 until 1856[5] |
| Herodsfoot | (and Herodscombe) North Herodsfoot Mine | Argentiferous Galena Cerusite Pyromorphite (with Sphalerite Baryte Stibnite Chalcopyrite Bournonite Tetrahedrite | Lode up to 1.22 m wide dips from 70°W[8] to E dip[9] | 19 316 tons 71% lead ore (1852–1883) 625 897 oz silver (1881–1884) 17 tons 6% copper (1881–1884)[2] 92 tons 52% lead 673 oz silver (1881–1882) 5 tons wofram (1881) | worked sporadically for 300 years[10] working in early 19th century, reopened in 1844 |

**Table 2** *continued*

| Mine/Location | Associated mines | Minerals | Notes | Production | Working period |
|---|---|---|---|---|---|
| Wheal Wray and Ludcott | North Trelawney | Earthy iron ore Pyrite Sphalerite Chalcopyrite Argentiferous Galena Silver Pyrargyrite Proustite Argentite | Western lode dips 70° E, eastern lode dips steeply west. 0.6–0.9 m wide | North Trelawney — 327 tons Lead 4014 oz silver (1852–1861)[2] Wrey Consols — 5435 tons 69% lead ore 107 020 oz silver (1853–1862) 3108 tons 70% lead ore 105 804 oz silver (1856–1862)[2] 12 tons silver ore 69 tons lead ore (1861) 306 tons silver ore (1864–1864)[8] Ludcott and Wrey 1870 tons 67% lead ore 81 580 oz silver, 87 tons silver-rich lead ore (1863–1866)[4] | Approx. 1850 to 1866 |
| Wheal Treweatha | | | 0.1–0.3 m wide lode dipping E with caunter lode | 4369 tons 67.5% lead ore 145 376 oz silver (1853–1872)[2] | |
| Wheal Trelawney | Wheal Trehane, Wheal Hony and Trelawney | Argentiferous Galena Fluorite Chalcopyrite Arsenopyrite Pyrite Baryte | 0.6–1.8 m-wide lode dipping 60–80°E<br><br>Dumps worked for arsenic, silver-lead and zinc (1900–1902) and for fluorspar and Barytes in 1919 | 24 653 tons 72% lead ore 920 tons pyrite 7230 tons Arsenopyrite 460 tons arsenic (1845–1871) 709 272 oz silver (1852–1857) As Hony and Trelawney — 88 tons 68% lead ore (1883–1884) Trehane — 4531 tons 66% lead ore (1847–1857) 79 649 oz silver (1852–1857)[2] | Approx. 1845 to 1884, 1900–1902 and 1919 |
| Wheal Mary Ann | | Fluorite Siderite Pyrite Argentiferous Galena | 0–1.2 m-wide lode dipping 79°E | 29 600 tons 69% lead ore (1848–1875) 1 125 130 oz silver (1854–1875)[4] | |
| Bere Alston–Bere Ferrers | Bere Ferris | | | 1400 tons lead ore 76 000 oz silver (1788–1795)[5] 10 tons black tine (1852–1913) 205 tons copper ore (1821–1913) 25 670 tons lead ore 534 010 oz silver (1852–1913)[11] | 13th century to 1913 |

| | | | |
|---|---|---|---|
| 1 De la Beche (1839) | 4 Dines (1956) | 7 Booker (1967) | 10 Stephens (1932) |
| 2 Beer (1988) | 5 Hamilton Jenkin (1974) | 8 Collins (1904) | 11 MacAlister (1921) |
| 3 Hamilton Jenkin (1967) | 6 Barton (1964) | 9 Giles (1865) | |

Drainage within the district is predominantly in a southerly direction by means of the rivers Tamar, Tavy, Seaton and Looe. Mean annual precipitation increases from 1000 mm along the coast to 1600 mm over the highest ground. Evapotranspiration averages 555 mm/a, with the highest values occurring along the coast.

Although there are no major aquifers in the district, there are currently 143 different sources licensed to abstract a total of 262 Ml/a of groundwater. The major part of this water (81%) is used for agriculture including spray irrigation (Table 4).

Yields from most of the formations within the district are generally low, one hour bailer tests rarely producing more than 0.5 to 1.0 l/s, from typical 150 mm diameter and 20 to 50 m deep boreholes. Small springs occur throughout the area where rocks of different permeabilities abut; some are associated with lithological or faulted boundaries shown on the map (e.g. north of Plymouth several streams commence as springs issuing from the boundary of the Saltash and Torpoint formations and the dolerite dykes commonly give rise to springs). Typical analyses of groundwaters from the different formations are given in Table 5.

## Hydrogeological characteristics of the principal aquifers

### DEVONIAN ROCKS

Devonian rocks, comprising slaty mudstone, siltstone, sandstone and limestone, underlie most of the district

**Table 3** Building stones available in disused quarries in the district.

| Lithology and colour | Quarry name and location | | Geological formation |
|---|---|---|---|
| Green slaty mudstone | Pentillie | 4122 6428 | Tavy Formation |
| Buff-weathering sandstone | Trehills Plantation | 4790 6199 | Trehills Sandstone Member (Tavy Formation) |
| | Dunstan | 3791 6637 | St Mellion Formation |
| | Crocadon | 3921 6575 | St Mellion Formation |
| | Nr. Lower Clicker | 2875 6063 | Staddon Formation |
| | Landlooe Bridge | 2490 5957 | Staddon Formation |
| | Hendergulling | 2255 5305 | Upper Longsands Sandstone Member (Bovisand Formation) |
| | Tor Lane | 4110 5841 | Wearde Sandstone Member (Saltash Formation) |
| Buff-weathering sandstone and dark grey slaty mudstone | Bake Wood | 3065 5780 | Bovisand Formation |
| | St Nonna's Well | 2237 5624 | Bovisand Formation |
| Dark blue-grey slaty mudstone | Burraton | 4105 6702 | Burraton Formation |
| | Trehunsey Barton | 3052 6556 | Saltash Formation |
| | Hepwell Bridge | 3032 6424 | Saltash Formation |
| | Kilquite | 3330 6185 | Saltash Formation |
| | Bokenna Wood | 2053 6672 | Saltash Formation |
| | Treskelly | 3345 5785 | Saltash Formation |
| | Craggs | 3609 5872 | Saltash Formation |
| | Hobb Park Farm | 2235 5661 | Bovisand Formation |
| | Lantoom | 2245 6490 | Saltash Formation |
| Limestone | Underways | 3420 5674 | Saltash Formation |
| Purple and green slaty mudstone | Wivelscombe | 3992 5718 | Torpoint Formation |
| Black chert | Amy Down | 3615 6664 | Newton Chert Formation |
| | Pillaton | 3715 6430 | Newton Chert Formation |
| | North Down Wood | 3850 6375 | Newton Chert Formation |
| Basalt | Heskyn | 3445 5953 | |
| | Trevethick Ball | 3035 6103 | |
| | Tilland Wood | 3315 6231 | |
| Basalt /Dolerite | Lynher River valley between Drillers Quarry 384 599 and Grove 376 574 | | |
| | Notter Bridge | 3840 6100 | |
| Dolerite | Wilton | 3182 5926 | |
| | Lambest | 3039 6347 | |
| | Hatt | 3968 6255 | |
| Picrite | Clicker Tor | 286 613 | |
| Hyaloclastite | Sawdey's Rock | 3635 6092 | |

and supply 90 per cent of the groundwater used in the area from fractures. Yields from individual sources within the Devonian rocks are small, generally less than 1.0 l/s for boreholes although a yield of up to 7.5 l/s has been recorded for a 2 m-diameter well.

The **Plymouth Limestone Formation** crops over a limited area in the south-east of the district around Plymouth and extends eastwards into the adjoining (Ivybridge) district where its main outcrop area occurs. The limestone is extensively recrystallised and possesses little primary porosity. Its water-bearing capacity depends on a well-developed series of fractures and bedding planes which have been enlarged by solution and endow the formation with a system of fissures and caverns. Borehole yields vary widely depending on whether water-bearing fissure systems have been encountered. Historically some of the better-yielding wells were

**Table 4**  Quantity of groundwater licensed to be abstracted from the Plymouth district in April 1993 (data supplied by South Western Region, National Rivers Authority).

| | Agriculture (m³/a) | Spray irrigation (m³/a) | Private water supply (m³/a) | Industrial processing (m³/a) | Total (m³/a) |
|---|---|---|---|---|---|
| Superficial deposits | (3) 3816 | (1) 2464 | (2) 497 | – | (6) 6777 |
| St Mellion Formation | (9) 11 033 | (1) 6818 | (1) 2045 | – | (11) 19 896 |
| Tavy Formation | (29) 43 647 | (3) 16 764 | (3) 1494 | – | (35) 61 905 |
| Torpoint Formation | (7) 6131 | – | (1) 17 528 | – | (8) 23 659 |
| Saltash Formation | (41) 60 779 | (1) 2054 | (6) 5299 | (3) 941 | (51) 69 073 |
| Staddon Formation | (3) 7187 | – | (6) 9275 | (1) 6900 | (10) 23 362 |
| Bovisand Formation | (18) 31 677 | – | (2) 435 | – | (20) 32 112 |
| Whitsand Bay Formation | (10) 17 194 | (1) 2864 | (7) 5133 | – | (18) 25 191 |
| **Total** | (120) 181 464 | (7) 30 964 | (28) 41 706 | (4) 7841 | (159) 261 975 |

Numbers in brackets refer to number of abstraction licences

those located along the faulted junction of the limestone with the Devonian slate. The limestone was exploited during the nineteenth and early twentieth centuries for industrial and brewery supplies. However due to pollution and saline intrusion there are no sources currently licensed for abstraction in this district. Farther east, some older sources are still operational and boreholes have been sunk recently to supply quarrying and cement manufacturing needs (GAPS, 1985). The limestone responds rapidly to recharge events as its storage capacity is limited. It also discharges groundwater rapidly into the adjacent slate formations and into Plymouth Sound via a series of submarine springs. Yields from boreholes have been as high as 13.9 l/s, although the water obtained is commonly brackish (approximately 900 mg/l chloride ion concentration) or salty (see Table 5). The water table in the limestone is generally discontinuous and controlled by the fracture systems and solution cavities. Caverns have been mapped both along the coast and inland associated with the fault zones, and brackish to saline groundwater bodies in the caverns present a potential hazard to foundations (GAPS,1985).

CARBONIFEROUS ROCKS

Carboniferous rocks are present in the north of the district. They are relatively unimportant for water supply due to their small area of crop. The **St Mellion Formation** is utilised for small agricultural and private supplies from boreholes, shallow wells, adits and springs.

PERMIAN ROCKS

Permian rocks occur only in the Kingsand area and they are of no hydrogeological significance due to their very small area of outcrop and proximity to the sea.

Throughout the district, groundwater occurs in joints and fissures, and yields are dependent upon the size of the fissure system a borehole intersects. Yields from the slates and grits are generally only sustainable for short periods of time, and may decrease in dry weather. The interbedded volcanic rocks and dolerite also yield small supplies where they are unweathered and fractured. Where they are weathered, however, the resultant clays are of low permeability. The groundwaters have short residence times in the aquifers, and are often similar in composition to surface waters with low total dissolved solids contents; they may be acid (pH <5.5). Sodium and chloride ion concentrations increase towards the coast. Locally, nitrate concentrations are elevated due to agricultural practice and may cause problems for domestic supplies.

Waters from the Plymouth Limestone Formation are hard, with pH values of about 7.5, bicarbonate ion concentrations of up to 500 mg/l and calcium up to 90 mg/l. The entire crop of the limestone aquifer is within 500 m of the coast and in hydraulic contact with the sea. Therefore most sources are now contaminated by sea water with increased chloride ion concentrations caused by saline intrusion occurring as a result of groundwater abstraction which has reversed the direction of the natural hydraulic gradient. Tidal fluctuations in groundwater levels have been observed in wells at Cattewater [497 537] just to the east of the district.

## AQUIFER VULNERABILITY AND POLLUTION

Rapid rates of flow through the highly fissured Plymouth Limestone Formation aquifer means that the aquifer is susceptible to contamination from surface waters, cesspits and leaky sewers. This is an historical problem, as Inglis (1877) reported that the houses sited on the Plymouth Limestone, used caverns within the limestone beneath the area as cesspits. Moreover, the Plymouth City sewer system was badly damaged by bombs during the war with consequent leakage of raw sewage into the aquifer. Groundwater in the Plymouth Limestone Formation is now so contaminated that it is no longer considered as a resource in this district.

**Table 5**   Typical analysis of groundwater in the district.

| Location | Talland* | Tredinnick† | Maker Heights* | Stonehouse* | Stonehouse‡ | Polkinghorne Brewery§ | Valletort Mills Plymouth† | Coldwind† |
|---|---|---|---|---|---|---|---|---|
| National Grid Reference | SX 2363 5127 | SX 236 571 | SX 4390 5163 | SX 4637 5436 | SX 4638 5435 | SX 45 | SX 4666 5421 | SX 208 658 |
| Type of source | Spring | Borehole | Borehole | Borehole | Borehole | Borehole | Borehole | Borehole |
| Aquifer | Whitsand Bay Formation | Bovisand Formation | Staddon Grit Formation | Plymouth Limestone Formation | Plymouth Limestone Formation | Plymouth Limestone Formation | Plymouth Limestone Formation | Plymouth Limestone Formation |
| Date of analysis | 21.10.1964 | 17.09.1992 | 16.10.1916 | 12.1941 | ? 1985 | pre 1877 | 28.09.1953 | 17.09.1992 |
| pH | | 5.0+ | | | 6.75 | | 7.3 | 5.2+ |
| Electricity conductivity (µmhos/cm) | 495 | 179 | | | 540 | | 3400 | 169 |
| Total dissolved solids (mg/l) | 250 | 59 | | 663 | | 771 | 2450 | |
| Total hardness (mg/l $CaCO_3$) | 170 | | 150 | | | | 700 | 53.5 |
| Bicarbonate (mg/l $HCO_3$) | 146 | | | 215 | 165 | 235 | 354 | |
| Sulphate (mg/l $SO_4^{2-}$) | 48 | 13 | | 89 | 25.0 | 248 | 263 | 13 |
| Chloride (mg/l $Cl^-$) | 60 | 23 | 30 | 164 | 20 | 134 | 900 | 21 |
| Nitrate (mg/l $NO_3.N$) | | 1.39 | Nil | 18.7 | 1.5 | 3.9 | 5.2 | 6.09 |
| Fluoride (mg/l $F^-$) | | 0.174 | | | | | | 0.072 |
| Calcium (mg/l $Ca^{2+}$) | | 8.0 | | 88 | 50.2 | 177 | 150 | 5.0 |
| Magnesium (mg/l $Mg^{2+}$) | | 9.5 | | 28 | 13.1 | 20.5 | | 10.0 |
| Sodium (mg/l $Na^+$) | | 14 | | 97 | 24 | 58 | | 10 |
| Potassium (mg/l $K^+$) | | 1.2 | | 11 | 4.2 | | | 0.5 |
| Iron (mg/l Fe) | | <0.05 | | Nil | 0.5 | | 1.6 | <0.05 |
| Manganese (mg/l Mn) | | 0.67 | | | | | | 0.0005 |

\*   National Well Record Collection
†   Environment Agency South Western Region
‡   GAPS, 1985
§   Inglis, 1877
+   Field value

In areas outside the outcrop of the Plymouth Limestone Formation there are two main types of groundwater pollution, from diffuse and from point sources. Diffuse pollution is caused largely by the application of fertilisers in the district leading to a rise in nitrate concentrations in groundwaters. Locally this is a problem in the shallow, fractured aquifers within the Devonian and Carboniferous rocks of the area.

Point sources of pollution such as landfills, storage tanks and some industrial processes and damaged sewers also represent a threat to groundwater quality. Although the Devonian and Carboniferous rocks of the district are generally of low permeability, point sources of pollution can present a hazard to groundwater quality, although the extent of such pollution in likely to be localised. Landfills of non-inert material present potential groundwater pollution problems unless they have been adequately lined. The leachate from unlined tips is capable of percolating through fissures where the Devonian rocks are heavily fractured and jointed, locally contaminating groundwater and, if the site is positioned on a slope above a water course, possibly forming a source of contamination of surface waters through surface and subfluvial springs. Of some concern in this regard are the pre-1945 rubbish tips in Moon Cove, Devonport [4496 5539], Honicknowle [4660 5855] and Ernesettle [445 599] (GAPS, 1985) and the widespread small abandoned building-stone and field-stone quarries that have been commonly used for the disposal of agricultural waste.

Some industrial processes, particularly those of Victorian and earlier industries such as gas and coke production, chemical works and tanning are potential sources of localised contamination of groundwater. Although there was apparently little industrial development in the district in Victorian times the topographical survey of 1854–57 (Ordnance Survey) shows a gas and coke works [4525 5596] at Keyham in Plymouth together with chemical works adjacent to the dockyard in western Plymouth [450 570], adjacent to Millbrook Lake [4375 5274] and at Torpoint [4405 5488]. The 1881–82 topographical survey (Ordnance Survey) also identified the presence of a small gas works at Liskeard [2508 6415] immediately adjacent to a tannery [2511 6420].

## DERELICT GROUND

Apart from industrial sites within urban Plymouth, much of the derelict land in the district has been derived from mineral workings. A significant amount of this land has been affected by quarrying operations but it is the sites of metalliferous mines which give rise to most concern (Cornwall County Council, 1996). Although little of this land lies within the area of potential growth of the larger urban centres of Plymouth or Liskeard, many of the sites lie on the outskirts of small villages and may be required for development at some time in the future.

Detailed information regarding the workings of the derelict mines in the district, essential for any assessment of potential development or ground stability, can be obtained from the extensive archive maintained by the BGS.

The derelict mine sites are characterised by underground workings and shafts (described in *Undermining* below), by waste tips of rock and mineral debris (mine dumps), and in many cases by accumulations of crushed and processed mine waste (tailings) in adjacent streams.

The clustering of derelict mines along particular mineral lodes in the district has resulted in **mine dumps** being grouped together. The main clusters of mine dumps are present between Menheniot [288 628] and Trebeigh Wood [296 668], between Weir Quay [434 647] and Bere Alston [439 666] and along the West Looe River at Herodsfoot [214 604]. Apart from these main clusters there are also a number of isolated mine dumps within the district, associated with small mining operations (Figure 4). These mine dumps contain a proportion of metallic ores excavated but not recovered during the mining operations. In the district, these ores are commonly those of iron, lead, zinc, copper, arsenic, silver and barium with possible traces of cadmium. These metallic ores represent a contamination hazard especially if they are leached into surface or ground waters (ENDS, 1992). Because of the long period during which many of these dumps have been exposed to weathering and leaching since the mines were abandoned around the turn of the century, much of the available contamination in the surface parts of the dumps has now been washed away. The potential for contamination of ground and surface waters does, however, still exist where the dumps have been disturbed. The mine dumps at Wheal Mary Ann, to the north of Menheniot, have undergone at least two periods of disturbance. Shortly after underground working at the mine ceased the mine dumps were reprocessed to recover further amounts of lead and silver. Not only did this cause the dumps to be excavated but the reprocessing operation washed additional volumes of tailings into the stream north and west of Menheniot, and thence into the River Seaton. This period of reprocessing continued probably until as late as the 1930s. Some of the mine dumps were then reprocessed again in the 1980s to recover fluorspar.

Almost all of the mining operations in the district crushed and washed the ore, using the local streams, to produce mineral concentrates. During these operations considerable quantities of comminuted sand-grade material (**mine tailings**), containing significant quantities of the same metallic ores as are present in the dumps, were washed into the adjacent streams and rivers. The pollution plumes in the river sediments down stream from these operations are still apparent and concentrations of copper, lead and zinc are apparent in geochemical surveys of the drainage downstream from the principle mine workings (IGS, 1981). Arsenic and cadmium were not analysed in this survey of stream sediments, but it is likely that they are also present as components of the mine tailings. In a similar manner to the dump material, the upper parts of this material have been leached by the stream flow since mineral processing operations ceased around the turn of the century. This has largely removed the contaminants from the surface layers of the alluvium in the alluvial flats and the stream bed itself. Disturbance of the alluvium, however, can rapidly cause pollution of the stream waters by exposure of the deeper, unleached layers of mine tailings. The stream valleys with the most potential for this type of pollution hazard in the district are: the River Tiddy, which shows elevated levels of copper, lead and zinc in its sediments from the northern boundary of the district at least as far south as Tideford [348 595]: the River Seaton, which shows similarly high levels of these metals throughout its course in the district, derived from the mines to the north: and the West Looe River which shows high levels of copper, zinc and lead in its sediments from Herodsfoot [214 606] at least as far south as Sowden's Bridge [230 555].

The **waste tips** within the district can be divided into two groups: those in which domestic and industrial waste has been deposited, and those in which constructional (both building and excavation) wastes have been deposited. The principal domestic and industrial waste tip in the district is at Holwood quarry [3609 6309] where waste has been tipped into a disused dolerite quarry on a ridge above the Lynher River. This tip ceased to be actively used only recently. Older sites for domestic and industrial waste disposal are on the alluvium at Lamerdon Mill [2506 6114] and at Honicknowle [4660 5855], although the tipping of refuse at Moon Cove [4496 5539] in Plymouth, during the 19th century, is reported by GAPS (1985).

A number of sites in the district have been used for the tipping of inert constructional wastes. Most of these sites are valley infilling such as the disused sites at Saltmill Creek [427 596], Thanckes Lake [434 557] and Carkeel [413 605]. The currently active site at Roodscroft Farm [395 614] is licensed for Category A and limited Category B wastes which are being used to infill a minor valley.

## UNDERMINING

The metalliferous lodes excavated by mining within the district are all relatively thin, steeply dipping veins which were excavated using underground mining techniques. The main feature of the mining methods used is the deep vertical or near-vertical shafts, either on or adjacent to the mineral lodes. They were used for access, ore extraction and drainage pumping. The shafts connected into a series of galleries (levels) extending along the mineral lode, and the lode was extracted by cutting out material (stoping) between the levels. This process generally leaves large underground voids (stopes) that may be several metres in width and hundreds of metres in depth, extending for considerable distances along the lode structure.

The mines were drained by pumping water from the base of the mine. To avoid the cost of raising the water to the ground surface a tunnel (drainage adit) was constructed to connect the pump shaft with the lowest level of an adjacent river valley. In places, these adits ran along the lode itself and served a dual purpose, to exploit the lode and to drain the mine. In other mines the adit took the shortest route to an adjacent valley or, more rarely, was routed so as to explore for additional mineral lodes. In the latter case, the adit could extend for considerable

distances away from the mine itself. An adit which is probably of this type extends from Treweatha Mine oblique to the mineral lodes to exit adjacent to the road at Trehunsey Bridge [6500 2950].

The modern undermining hazards associated with this method of mineral extraction are threefold.

1. Hazards related to the stability of the shafts themselves. There may be a small amount of local collapse around the tops of open shafts where the rock is most fractured and weathered, but generally shafts which have not been infilled are relatively stable.

Shafts which have not been capped, but have been infilled by tipping waste into them are potentially unstable. Many of the access shafts originally had wooden platforms or cross beams in the shafts to support the access ladders. Where tipped debris has become lodged on these platforms and 'choked' the shaft, the infilled surface may appear stable until the wooden support rots or gives way under loading. In this situation the shaft filling can collapse catastrophically. A collapse in 1979 over a mine shaft (Engine Shaft of Wheal Whitleigh) beneath a housing estate in the northern Plymouth suburb of Whitleigh [483 596] may have been caused in this manner.

2. The instability of the workings themselves forms a hazard to the development of the surface areas above the workings. Although a large proportion of the workings are at a considerable depth below the surface and their instability would be unlikely to affect the surface, some parts of the workings are at shallow depths. Adits in particular generally lie at relatively shallow levels, particularly near their portals where they intersect the valley sides to allow the egress of the drainage waters. Because adits do not necessarily follow the course of the mineral lodes, the subsidence hazard associated with them may exist outside the area occupied by the mine site itself. An example of this potential hazard is Treweatha Mine [291 654], where the mineral workings have a north–south trend but the drainage adit trends north-west–south-east with its portal over 600 m away from the mine site.

The stoping of the steeply dipping mineral lodes may also extend upwards to shallow levels, in the case of Wheal Trelawney [289 639] above the level of the adit itself, and form a significant subsidence hazard, although this will be limited to a relatively narrow zone along the course of the mineral lode. At South Tamar Consols [437 646], shallow stoping to within 30 m below the bed of the River Tamar caused the river to break into the workings and flood the mine in 1856. At South Hooe Mine [424 656] the mineral lode between the levels has been completely taken out forming a void over 300 m deep with its top less than 25 m below the bed of the River Tamar resulting in a significant risk of surface subsidence. Similarly large voids have been formed by stoping at both Wheal Mary Ann [287 635] and Wheal Trelawney [289 639].

3. A further hazard concerns the mine waters draining from the workings. When underground mining activity ceased at around the turn of the century, pumping of

the drainage waters also ceased. The mine workings were then allowed to fill with water until they over-flowed along the drainage adits. This undoubtedly caused pollution of the watercourses below the adit portals in a manner similar to that which occurred in west Cornwall following the overflowing of Wheal Jane in 1992, this pollution was relatively short-lived. If, however, the drainage adits were to become blocked by the collapse of the shallow tunnels themselves or by some other means, the mine workings may then fill completely. This presents several environmental hazards:

- As the mine workings fill with water the residual metalliferous ores in the higher workings, which have been oxidising for more than a century, will be submerged. Contact between the acid mine waters and these minerals will cause serious contamination of the waters themselves and these in turn will cause serious pollution of watercourses, of the same type as occurred at Wheal Jane, if they are released. Furthermore the pressure of water dammed behind an adit blockage may be sufficient to remove the blockage, releasing the contaminated mine waters catastrophically.

- Collapses of levels or of material choking shafts into water-filled workings can cause water and other mine debris to be ejected from other shafts or cause surface collapses above shallow level stopes. In 1981 this situation occurred at Yarner Wood mine, south-east of Dartmoor when debris in a shaft collapsed into shallow workings flooded by the blockage of the drainage adit. The resulting overpressuring of the water in the mine caused the ground over the stope to be blown out and a considerable quantity of rock, water and mine timbers was ejected from the mine.

- The filling of mine workings by water can also initiate collapses of the looser near surface material at the tops of shafts. In 1992 the collapse of Michael's Shaft of Old Gunnislake Mine to the north of the district at Gunnislake, beneath a housing estate, was apparently due to blockages in the mine drainage, causing water levels within the mine to rise and wash out loose material below the shaft cap, allowing it to collapse into the shaft.

It is therefore most important that any development of the mine sites in the district is accompanied by monitoring and maintenance of the drainage adits and that the development does not lead to the blockage of such adits in any way. Further informa-tion relating to the location of these adits can be obtained from BGS.

## MADE GROUND/FILL

With the exception of urban Plymouth much of the made ground within the district takes the form of mine tips and waste tips (see above), quarry spoil tips and 'cut

and fill' associated with railway and major trunk road cuttings and embankments.

Quarry tips, associated with small building and walling stone quarries, are widespread but generally small. A few of the larger quarries have more substantial tips, such as those at Clicker Tor quarry [2860 6135] and the Brickpits at Inswork [4245 5266, 4325 5238, 4355 5284]. The sub-stantial quarry tips at Lean quarry [2650 6140] are also associated with the disposal of excavated waste from con-struction and other sites (see above). The substantial quarry tips at Holwood quarry [3609 6309] are composed entirely of rock waste and can be distinguished from the tipping of industrial and domestic waste which took place within the quarry itself (see above).

A substantial amount of made ground created by 'cut and fill' techniques is associated with routing of the A38 trunk road, together with various abandoned 'meanders' as the route is improved and villages along it are bypassed, as it cuts east–west across the predominantly southerly drainage of the district. Much of the rest of the 'cut and fill' made ground is associated with the east–west route of the main-line Plymouth to Penzance railway line. A small number of embankments are present on the Liskeard–Looe branch-line, although this runs for much of its length along the valley of the East Looe River. Some railway embankments are also present along the route of the now abandoned Caradon–Liskeard railway.

Within the Plymouth urban area, large areas of made ground are associated with the creation of the Dockyard on the alluvial flats on the eastern bank of the River Tamar. Much of the material excavated from the dock basins was tipped onto the estuarine alluvium of the Tamar to form the dockside itself (GAPS, 1985). Other areas of made ground along the creeks in the estuary frontage of Plymouth at Western Mill [4490 5715], Ernesettle [445 599], Stonehouse Lake [464 548], Millbay [4665 5400] and Sutton Harbour [485 545], were formed by damming sections of the creeks and leaving the alluvium to accumulate. At Sutton Harbour, Weston Mill and Ernesettle, layers of rubble or household and industrial waste were tipped on top of this alluvium. At Ernesettle, these layers were augmented by the tipping of rock waste from the excavation of the adjacent under-ground munitions storage tunnels.

Large areas of made ground in urban Plymouth are related to the intense bombing of the city during the 1939–45 war. Much of the city centre was destroyed at this time and the area is now mantled in a variable thickness of building rubble, although small patches of made ground, not definable on the geological map, infilling bomb craters are present all over the city. Some of the valleys within the city limits, as at Honicknowle [4660 5855], were partially infilled by bomb rubble.

## FOUNDATION CONDITIONS

The suitability of bedrock and superficial materials for foundations depends on their geotechnical properties. The various conditions which affect the engineering properties of the bedrock and drift (superficial) deposits, principally in and around urban Plymouth, are examined below. Geological structures, including those causing bulk anisotropy such as slaty cleavage and crenulation cleavage, together with those generating discrete discontinuities such as jointing and faulting are described with specific reference to the Plymouth Limestone Formation below and in more general terms in Chapter 10. Slope stability (see *Coastal slope failure*), mining cavities (see *Undermining*), natural caverns (see *Solution cavities*), the mineralogy of the bedrock and the degree of weathering, all have a bearing on the engineering properties of the bedrock.

Detailed information regarding the disposition and orientation of discontinuities within the rock, together with an extensive archive of geological and geotechnical information, of prime importance in evaluating the suitability of a site for development or redevelopment, is available from BGS.

## Bedrock

### WEATHERING IN SLATY MUDSTONE OF THE TORPOINT AND SALTASH FORMATIONS

The Torpoint and Saltash formations form a laterally extensive outcrop from the eastern to the western boundary of the district, and in width extending from central Plymouth in the south to Landulph [431 615] in the north. Because these slaty mudstone rocks lay beyond the edge of the Pleistocene ice sheets, prolonged exposure to weathering and periglacial processes has resulted in the development of a softer surface layer known locally as 'shillet'. This material has formed in situ as a residual soil which, although generally thin, is of variable thickness. The thickness and properties of this 'shillet' and the structural weaknesses in the weathered and fresh slaty mudstone bedrock are important engineering geological considerations for engineers and planners. The weathering of these rocks has resulted in an increase of porosity and decreases in unit weight (bulk density), shear strength and stiffness.

A summary of weathering grades, together with their geotechnical properties, for the Torpoint and Saltash formations in the Plymouth city area is presented in Table 6. There is not a large difference in engineering behaviour between these weathering grades, and there can be some geotechnical variation within each grade. Because of this, the descriptive interpretations of these weathering grades cannot be used as the sole guide to material behaviour. In the slaty mudstone lithologies of the Torpoint and Saltash formations in the Plymouth city area the most important engineering boundaries lie between grades V, IV and III where the weathered material changes from a soft to firm soil-like material of low to medium plasticity to a very weak to weak rock, although in places medium strong to strong rock strength categories may be present.

Groundwater within the slaty mudstone rocks of the district lies at deeper levels on hilltops and shallower levels in valleys as the groundwater flows downslope. The effective stresses and strengths of the rocks will therefore tend to be lower in the valleys. Below 10 m depth the apparent lack of groundwater in test borehole observations suggests that the formations are watertight, and that there is, therefore, a significant change in hydraulic conductivity below this depth. Packer tests carried out at various depth up to 50 m through weathering grades III to I, suggest that the hydraulic conductivity varies from low to very low (between 0.1 and $25 \times 10^{-7}$ m/sec).

Shear box tests on wet surfaces of both joints and cleavage surfaces have been carried out on grade II–III rocks from a number of geotechnical boreholes in the Plymouth city area, at the effective normal stresses shown in Table 6. Although the condition of the surfaces themselves was not described in these tests, the tangent values for the Effective Cohesion and the Effective Friction Angle were apparently strongly stress dependent. At low normal stresses, (as in a cutting slope or excavation) dilatancy provided a significant strength component to peak friction angles (see Table 6). Residual values for the Effective Cohesion were 20 kPa at the same level of stress with an Effective Friction Angle of 26° (see Table 6). These values can be compared to back analysed results from slope failures in the nearby Whitsand Bay Formation (Dartmouth Group) at Fowey described by Fookes et al. (1977) where values of 3 kPa for the Effective Cohesion and 24° for the Effective Friction Angle were adopted for remedial design.

Figure 5 represents a generalised profile through the slaty mudstone of the Torpoint and Saltash formation showing the trend of weakening (R) with depth and the variation in Young's Modulus. Weathering zone depths in this figure represent a typical sequence through the slates down to a faintly weathered bed rock at 8.5 m. Values of R decrease to 0.1 between weathering grades II and III, suggesting that locked-in strain energy may be an engineering consideration in grades I and II, and this is supported by Littlejohn and Bruce (1979), who reported that the performance of some rock anchors in slates at HM dockyard, Devonport had been affected by high in-situ stresses in the rock. There is, however, little information on the strength and stiffness of the unweathered grade I slates.

There is some evidence that the weathering grades I to III may be strongly anisotropic. Littlejohn and Bruce (1979) quote an anisotropic index (the maximum intensity of preferred stiffness to a random distribution of that stiffness) for Young's Modulus ranging from 8 to 18 indicating that the results shown in Table 6 represent no more than a general guide to in-situ behaviour. This, however, emphasises the importance of correct assessment of the depth and stiffness of the less anisotropic grade IV material which would normally be the preferred foundation material for heavy or sensitive structures. Weathering grades V and VI provide the foundations for light structures such as pavements, housing and some industrial units in Plymouth. As discussed above, weakening theory suggests that the residual soils within these grades will have a wide range in porosity and possible secondary bonding. Each site may therefore be different and the summary of behaviour and predictive soil engineering is therefore problematic.

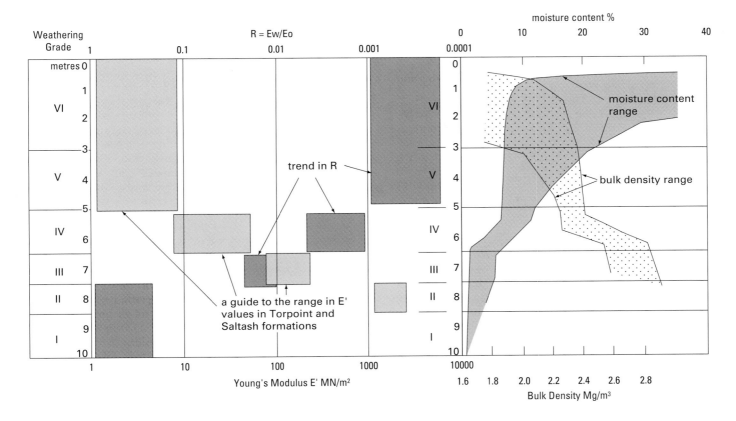

**Figure 5** A composite weathering profile in the Torpoint and Saltash formations: A guide to variations in stiffness, (E′), moisture content and bulk density with depth. Values of E′ for grades V and VI have been calculated from the oedometer coefficient of volume compressibility values assuming a Poisson's Ratio of 0.3.

R ratio of E weathered to a nominal unweathered value of E.

Some residual soil profiles in Plymouth may have thin zones with structured soils of comparatively high void ratio with the potential for loss of strength and large deformations.

SLOPE STABILITY IN DEVONIAN ROCKS IN URBAN PLYMOUTH

Discontinuities within the Plymouth Limestone Formation, and the slaty mudstone and volcanic rocks of the Torpoint and Saltash formations, together with long-term weathering, have exerted a strong influence on the geotechnical behaviour of the Devonian rocks of the Plymouth city area. The main discontinuities in the slaty mudstone and limestone forming the bedrock in urban Plymouth are bedding, cleavage, faults and joints. The orientations of these discontinuities are presented in Figure 6 and the general orientation of the major discontinuity planes summarised in Figure 7a.

The stability of slopes in the Devonian rocks of the Plymouth city area is dependent upon its orientation relative to the structural discontinuities in the bedrock (see Matheson, 1983, 1988). A planar discontinuity dipping out of a slope is likely to fail if it is inclined at a greater angle than the friction angle for that rock. This is also the case for the dip of the intersection lineation in

wedge failure. For toppling failure the basal release plane must dip out of the slope at less than the friction angle, otherwise plane failure would occur.

A guide to the expected friction angles for different rock types is presented in Table 7. These angles are for unweathered rock. Because discontinuities in the bedrock have acted as pathways for the deep penetration of the rock and within the slaty mudstone lithologies this has generated thin, often porous, seams of clay along planes parallel to the slaty cleavage (see Fookes et al., 1977) and in places parallel to the bedding in mudstone beds and laminae in the Plymouth Limestone Formation, the actual friction angles for weathered rocks in the Plymouth area may be much lower. Joints in the Devonian rocks in the Plymouth city area are, however, generally high angle and are therefore able to act as lateral release surfaces in all types of failure. The representation of the major natural discontinuities in the Plymouth city area in Figure 7a provides a generalised view of the orientations of these discontinuities. Only a selection of the discontinuities shown are likely to be represented at any particular locality, however, any local variations in their orientation would alter their interrelationships, changing the resultant failure modes. Even a small variation could, for

**Table 6**   Summary of weathering grades and geotechnical properties in the Torpoint and Saltash formations, Plymouth.

| Weathering grade | Diagnostic features | Characteristics Plymouth area | Boundary groups | Thickness (m) | LL % | PL % | Mc% (soil) | Particle size (mm) | Mc% (rock) | K x 10 m/sec | Bulk Density Mg/m³ |
|---|---|---|---|---|---|---|---|---|---|---|---|
| Residual Soil VI | Rock completely changed to a soil, rock fabric destroyed | Brown/red/orange soft to firm clayey silt with slate fragments | Depth to Rockhead ↓ | Torpoint and Saltash slates (0–10) | 28–57 | 8–24 | 8–36 | D80– (0.02–0.06), D50 (-0.0075–0.015), D20 (0.005–2.0) | | | 1.78–2.25 |
| Completely Weathered V | Rock is decomposed and in friable condition original fabric preserved | Brown clayey silts, fissured, many slate fragments, disturbed by periglacial processes | Weathered bedrock | Plympton sector (0.4–4.9), City central sector (0–5.15) | | | 8–20 | D80, (2.5–25.0), D50, (0.10–12.0) D20, (0.005–2.0) | | | 1.78–2.25 |
| Highly Weathered IV | Rock is discoloured discontinuities open, discoloured surfaces, alteration to rock penetrates deeply into fabric | Joints and cleavage open, faces lined with clay, some bands of limonite up to 100 mm wide | | Western sector (0–0.80), N central sector (1.4–6.72), N west sector (0.3–4.9) | Ultimate bearing capacity from plate loading tests 352–785 kN/m² | | | | 3.0–12.0 | 0.1–25 | 2.25–2.87 |
| Moderately Weathered III | Rock is discoloured, discontinuities are open, rock not friable | Joints and cleavage open, surfaces iron stained, lined with brown silty clay | | Northern sector, (0–2.0) | Ultimate bearing capacity from plate loading tests >1350 kN/m² | | | | 0.5–6.0 | | 2.54–2.87 |
| Slightly Weathered II | Open discontinuities with discoloured surfaces, intact rock not weaker than fresh rock | Joints and cleavage surfaces frequently iron stained | | All slates (0–6.72) | Peak shear Strength (Joints) $\sigma_n'$ / $c'_p$ / $\phi^1_p$: 0/0/80, 50/65/54, 150/100/44, 300/165/34, 450/290/19 | | | Residual Shear Strength (Joints) $c'_r$ / $\phi^1_r$: 0/55, 20/26, 35/19, 35/19, 35/19 | | | 2.54–2.87 |
| Faintly Weathered IB | Discolouration of only minor discontinuities | Joints and cleavage tight, some iron staining | | | (Cleavage) $\sigma_n'$ / $c'_p$ / $\phi^1_p$: 0/0/75, 50/45/51, 150/100/30, 300/135/24, 450/200/15 | | | (Cleavage) $c'_r$ / $\phi^1_r$: 0/43, 15/23, 35/9, 50/6, 70/2 | | | |
| Fresh IA | No discolouration or loss of strength | | Fresh bedrock | | (Shear box tests on wet specimens — tangents to failure envelope within normal effective stress range 0–450 kPa) | | | | | | |

Cu Undrained shear strength;   Mv Coefficient of volume compressibility;   Cv Coefficient of consolidation;   SPT (N) Standard penetration test;
LL Liquid limit;   PL Plastic limit;   Mc Moisture content;   D80, etc. % diameter passing;   c′ Effective cohesion;   $\phi_\mu$ Undrained friction angle;
φ′ Effective friction angle;   E′ Young's modulus;   K Hydraulic conductivity

$\sigma_n'$ effective normal stress;   $c'_r$ effective residual cohesion;   $c'_p$ effective peak cohesion;   $\phi'_r$ effective residual friction angle;   $\phi'_p$ effective peak friction angle

example, alter whether an intersection lineation was steep enough for wedge failure to occur.

Within its limitations the orientation data suggests that the minimum slope stability for **plane failure** is most likely to be encountered on south-facing slopes. If bedding, cleavage or one of the southerly dipping faults daylighting on such a slope dips more steeply than the

angle of friction for the rock, plane failure is likely to occur.

The orientations of design slopes in which probable plane failure out of the cut slope of each of the major discontinuities could occur are represented in Figure 7b. Plane failure is most likely in the Plymouth Limestone Formation, where bedding forms a major discontinuity.

**Table 6** *continued*

| Weathering grade | Point load index MN/m² | Undrained strength kN/m² | Allowable net bearing pressure | c' kN/m² | φ' | Mv m²/MN | Cv m²/year | SPT (N) | SO₃ % | Cl₂ % | pH |
|---|---|---|---|---|---|---|---|---|---|---|---|
| Residual Soil VI | | Lower values 28  11 Upper values 90  25 (partial saturation) | 80 kN/m² | 0 18 | 32 32 | 0.08–0.35 | 10–23 | 10–100 | 0.01–0.38 | 0.01–0.02 | 5.1–8 |
| Completely Weathered V | | Lower values 25  0 (m >32%) Upper values 235  0 (m <16%) (saturated) | 80 kN/m² | | | | | | | | |
| Highly Weathered IV | 2.0 | Unconfined compression strength Mn/m 1–12 | 100 kN/m² | E Laboratory × 10 MN/m²: Uniaxial \| Triaxial \| Geophysical | | | | E Field × 10 MN/m²: Plate load 0.03 \| Geophysical 27 (Poisson's ratio 0.27) | | | |
| Moderately Weathered III | 3.0 | 1–12 | 1500 kN/m² | | | | | Plate load 0.140 \| Geophysical 43 (Poisson's ratio 0.26) | | | |
| Slightly Weathered II | 4.5 | 1–>12 | 1500 kN/m² | Uniaxial 1.97 \| Triaxial 5.2 \| Geophysical 61.8 (anisotropy Index from 8 to 18 (mean value 11) | | | | Geophysical 53 (Poisson's ratio 0.27) | | | |
| Faintly Weathered IB | | >12 | 1500 kN/m² | | | | | | | | |
| Fresh IA | | >12 | 1500 kN/m² | | | | | | | | |

Plane failure is also probable in the Torpoint and Saltash formations beneath urban Plymouth, Torpoint and Saltash where cleavage dips are steep. In the Saltash Formation south of the Plymouth Limestone Formation, however, south-facing slopes are likely to be more stable because the cleavage is generally more gently inclined than the angle of friction. Degradation of the slaty mudstone to clay along cleavage planes, especially in the presence of water, will, however, locally decrease the angle of friction significantly. Plane failure could also occur on steep north- and north-west- facing slopes throughout the Plymouth city area by failure along the west-south-west-trending joint planes.

A general indication of probable **wedge failure** configurations for the Plymouth city area is presented in Figure 7c. Wedge failure at the intersections of the cleavage or the steeper elements of the bedding and the north-north-west-trending joints will, at least in part, merge with plane failure because they have similar orientations. Localised wedge failure in a south-easterly direction could occur at the intersection of the north-west-trending steep joints and the cleavage or the bedding,

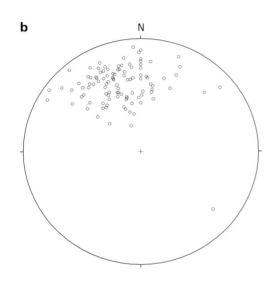

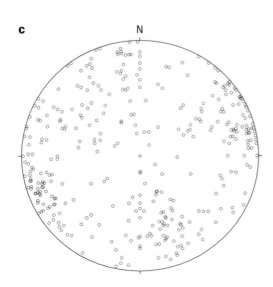

**Table 7** Guide to friction angles in different lithologies. Plane or wedge failure may occur where the dip of the relevant plane of intersection exceeds the friction angle (after Matheson, 1983).

|  | Rock type | Friction angle |
|---|---|---|
| Hard igneous | Granite, dolerite, basalt Porphyry | 35°–45° |
| Metamorphic | Gneiss, schist, slate | 30°–40° |
| Hard sedimentary | Limestone, sandstone conglomerate | 35°–45° |
| Soft sediment | Sandstone, coal, shale | 25°–35° |

principally in areas of the Plymouth Limestone Formation where the bedding is steeper (see Figure 7c). The intersection of north-north-west and north-west joints and faults and very steep east–west-trending joints could also produce south-easterly to south-south-easterly wedge failures in near-vertical cuttings. The intersection of cleavage and the near-vertical east-west joints lies above the possible friction angle and could therefore fail in an easterly direction in slaty mudstone beds where the cleavage is well developed. This type of failure is less likely in the Plymouth Limestone Formation where the cleavage is less intensely developed.

**Toppling failure** (see Figure 7d and e) on southerly dipping slopes is possible off shallowly dipping faults, bedding or cleavage with northerly dipping west-south-west-trending joints acting as rear release planes and north-north-west-trending joints acting as lateral release planes. The likelihood of this occurring, however, is not considered to be great because west-south-west-trending joints are not sufficiently closely spaced. Toppling failure is more likely to occur on steep north-facing slopes where west-south-west-trending joints dip northwards allowing blocks to detach from steeply dipping bedding or closely spaced steep cleavage planes. This form of toppling failure is illustrated at Mount Batten Point [4862 5324] where steep bedding and cleavage with average dips of 76° present a considerable toppling hazard on the steeply dipping north-facing slope below the tower.

More shallowly dipping north-facing slopes and most east- or west-facing slopes, perpendicular to the strike of most of the major discontinuities, are likely to be the most stable.

Instability in quarry faces in the slaty mudstone of the Torpoint and Saltash formations has been observed on cleavage planes dipping in excess of 40° and on bedding

**Figure 6** Discontinuity data for the Plymouth, Saltash and Liskeard area (Schmidt equal area projection).

a. Poles to bedding
b. Poles to cleavage (S1)
c. Poles to joints and faults

**Figure 7** Slope failure on discontinuities in the Plymouth, Saltash and Liskeard area.

a. Generalised discontinuities
b. Plane failure
c. Wedge failure
d. Toppling failure (south-facing slopes)
e. Toppling failure (north-facing slopes)

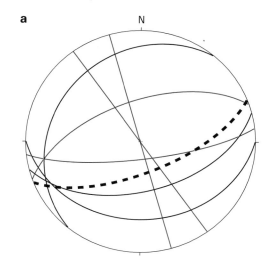

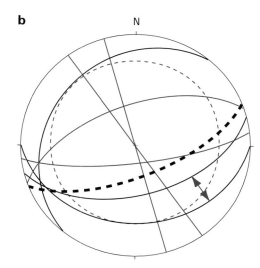

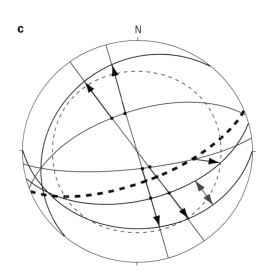

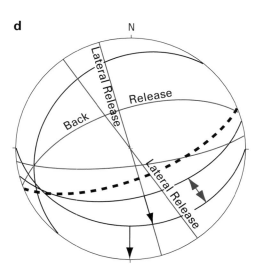

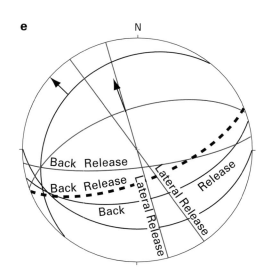

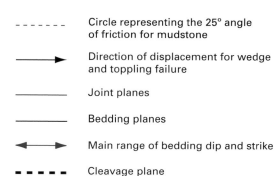

Circle representing the 25° angle of friction for mudstone

Direction of displacement for wedge and toppling failure

Joint planes

Bedding planes

Main range of bedding dip and strike

Cleavage plane

planes dipping more steeply than 66°. Steep quarry faces in these lithologies also show failure along joint planes. Instability in quarry faces in igneous intrusions within the Saltash Formation, where bedding and cleavage are absent, occurs along joints and faults with dips in excess of 30°. Outside the Plymouth city area Dearman (1991) estimated the maximum stable slope parallel to the strike of cleavage and bedding (approximately east–west) in unweathered lithologies as 48°, but this was decreased to only 25° in the upper parts of the slope where the slaty mudstone of the Whitsand Bay Formation was weathered.

In general, although hill slopes within the district are commonly steep, the steeper inclination of the cleavage which forms the dominant discontinuity within the upper parts of the weathering profile has inhibited plane failure within the slaty mudstone lithologies. Toppling of the slaty mudstone along the cleavage planes is, however, common forming a zone of 'hill creep' up to several metres thick below rockhead. This form of hill creep is generally absent within the more isotropic lithologies, such as massive sandstone, limestone and igneous intrusions.

Coastal slope failure

The coast within the district, even within the protected natural harbour of Plymouth Sound, is characterised by moderately high to high cliffs interspersed with a small number of narrow river valleys. All of these cliffs are currently subject to erosion by small-scale wedge failures and toppling, and parts of the coast are subject to larger-scale slope failures.

Two areas, one to the west of Rame Head in Whitsand Bay and another to the east of Looe, are affected by **landslip**. Between Wiggle Cliff [417 503] and Withnoe Common [403 517] numerous small landslips, individually with areas up to 25 000 square metres, are closely spaced. In this area the cliff section is in excess of 100 m high and is within 150 m of, and is subparallel to, the Rame Fault. Some of these landslips are suspended in the cliffs and comprise rock masses and debris toppled from higher levels, but others extend to beach level where the bases pass into a zone of fractured and deformed strata that indicate seaward displacement of the slipped mass by a zone of deformation rather than a discrete basal slip plane. Two of the larger slips [417 505 and 412 509] are spatulate in plan view, each extending to an apex some 200 m inland across the Rame Fault. The former of these two slips is currently active, with collapse extending landwards at the apex, having been observed to affect the coast road in the last year of the survey. The latter of these slips is at the centre of an established zone of weakness with Tertiary retreat cliff featuring cutting back towards the slip apex. This suggests that the locations of the slips may be determined by faults perpendicular to the Rame Fault. The sides of the slips have similar trends, east-north-east and north-north-east. The former is parallel to the strike of bedding and cleavage, and the latter to a prominent joint set in the district, suggesting that the slips are caused by large-scale wedge failure symmetric about north-east-trending faults.

The most substantial coastal landslip in the district lies between Millendreath Beach [2695 5405] and Seaton Beach [2965 5426]; a distance of nearly 2.75 km. At its widest the slip extends between 200 and 300 m back from the coast. It is a complex rotational slip formed within interbedded mudstone and sandstone of both the Whitsand Bay Formation and the Bovisand Formation. The structure forming the basal slip plane of the structure is not known. The strike of bedding and cleavage varies along the length of the slip. At the western end they trend parallel to the back-scarp of the slip and dip moderately to shallowly seawards. In the eastern half of the slip, however, the strike of bedding and cleavage becomes progressively oblique to the trend of the back-scarp until, at the eastern end of the slip, they are almost perpendicular. Cracks appearing in driveways at the rear of the slip at the date of survey indicate that movement of the landslip is currently taking place.

Between Portnadler Bay [242 516] and the western margin of the district the cliffs are formed of interbedded slaty mudstone and sandstone of the Whitsand Bay Formation. The steep northerly dip of bedding and cleavage along this stretch of high coastal cliffs with near-vertical and subhorizontal joints at a high angle to bedding has caused general and widespread **toppling failure** in the cliffs.

Solution effects in the Plymouth Limestone Formation

*Surface solution features*

The Plymouth Limestone Formation forms a low-lying limestone plateau in the southern part of urban Plymouth varying in height from 15 to 35 m above OD. Towards the ground surface the limestone is fractured into smaller and smaller blocks until these become dis-aggregated in the regolith. Solution effects on the surface of the limestone have caused the formation of a karstic surface on the bedrock. Because of the dominance of discontinuities in the solution of the limestone, many of these solutional depressions in the rockhead may be linear or angular rather than circular. Dolines associated with bedding in the limestone are generally wide and shallow, whereas those associated with near-vertical fault and joint planes may appear narrow at the surface but be very deep.

*Solution cavities*

Solution effects, although widespread throughout the Plymouth Limestone Formation, appear to be related to specific planar structural elements within it. Although most of the known solution cavities lie within the massively bedded limestone, a few caves are also present within the distinctly bedded Prince Rock Member of the Plymouth Limestone Formation (see Chapter 4). Solution has, however, taken place even within the thinly interbedded limestone and slaty mudstone of the Faraday Road Member where circumstances have permitted a pathway for groundwaters.

Both exposures in abandoned limestone quarries in the city area and boreholes within the massive limestone itself reveal that solution effects are strongly developed on those north-north-west-trending joints which have

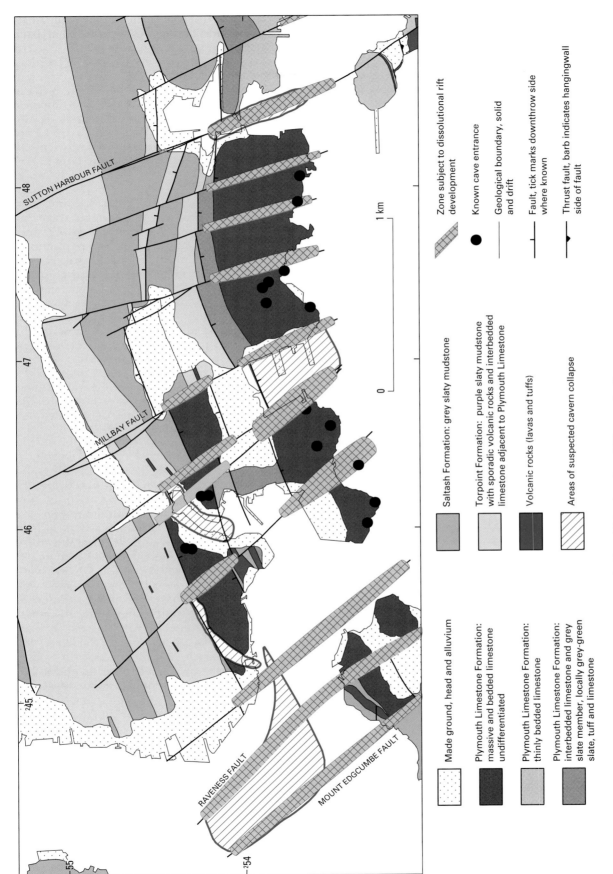

**Figure 8** Potential locations of solution caverns and rifts within the city of Plymouth. Cave entrance locations are from Jean (1984).

Made ground, head and alluvium

Plymouth Limestone Formation: massive and bedded limestone undifferentiated

Plymouth Limestone Formation: thinly bedded limestone

Plymouth Limestone Formation: interbedded limestone and grey slate member, locally grey-green slate, tuff and limestone

Saltash Formation: grey slaty mudstone

Torpoint Formation: purple slaty mudstone with sporadic volcanic rocks and interbedded limestone adjacent to Plymouth Limestone

Volcanic rocks (lavas and tuffs)

Areas of suspected cavern collapse

Zone subject to dissolutional rift development

Known cave entrance

Geological boundary, solid and drift

Fault, tick marks downthrow side where known

Thrust fault, barb indicates hangingwall side of fault

SUTTON HARBOUR FAULT

MILLBAY FAULT

RAVENESS FAULT

MOUNT EDGCUMBE FAULT

1 km

0

undergone some degree of fault displacement (see Figure 8). Solution cavities of this type take the form of parallel vertical rifts, a few centimetres to more than a metre in width, within the fault zone. The wider solution rifts have in places amalgamated with adjacent rifts to form substantial cavities (see Jean, 1984). Bedding planes abutting these rifts have also been exploited by solution, especially where the bedding is southerly dipping. These cavities are low, elongate, and irregular, or form subhorizontal phreatic tubes along the bedding strike. They commonly extend the rift cavities laterally, and interconnect them to form more extensive cave systems (Jean, 1984). Solution has also taken place along east-north-east-trending joints where these have a southerly dip.

This pattern of solution appears to be consistent with the current active southerly run-off of water southwards into The Sound at Plymouth from the slaty mudstone lithologies of the Torpoint and Saltash formations north of the limestone. The presence of cave entrances at sea-level is therefore consistent with their recent formation. However, a number of caves have entrances between 25 and 32 m above OD (Jean, 1984) suggesting at least one episode of formation during a period of higher sea level.

The presence of deeper level caves beneath Plymouth is suggested by a report from the Plymouth Area Police Diving Group who, in 1977, discovered a submarine cave system at approximately 30 m below OD in Firestone Bay [4648 5336]. Strong tidal current activity within this cave suggested that the system was relatively extensive. Discrete depressions in the floor of the buried submarine channel at Hamoaze (Eddies and Reynolds, 1988) extend down to approximately 40 m below OD and correspond with the position of north-west-trending faults along the western edge of the Plymouth Limestone; these depressions indicate the possibility that cave systems formed at this level. The presence of solution rifts in boreholes within the Plymouth Limestone extending down to 27 m below OD indicate that both deep and shallow cave systems may be interconnected along these rifts. The main breaches through the limestone outcrop in the Hamoaze [448 538], Stonehouse Creek [461 543], Millbay [4685 5385] and Sutton Harbour [484 540] may represent areas of cavern collapse on both the deep-level caves and the solutional rifts themselves (see Figure 8).

### Cavern infilling

The weathering deposits in the larger solution cavities in the limestone contain imported material in some places, consisting mostly of sand, quartz gravel and lithic clasts, either in pockets or beds, or enclosed in a matrix of limestone weathering residues. The limestone fragments and cobbles found in the cavities are generally collapse debris. This debris forms 'chokes' in some of the caves, potentially with voids below. In some circumstances water running into or through the caves can flush out the supporting matrix of the boulder chokes, causing them to collapse. The appearance of voids in excess of 3 m by 4 m beneath houses in Elliott Terrace on Plymouth Hoe

[4754 5394] in 1992 (Western Morning News, 23/1/92) is believed to have been caused by surface water from soak-aways flushing out cave sediments infilling the upper part of a void during the approximately hundred years since the houses were built.

Although substantial volumes of sediment infilling caves has not been observed within the district, adjacent to it, in the eastern part of Plymouth city, a cavity at least 20 m by 8 m by 5 m high filled with organic dark grey sandy clayey silt and red-brown clay was uncovered by quarrying operations at Pomphlett quarry [5120 5430].

### Solution effects on slope stability

Dissolution of limestone along the major discontinuities within the Plymouth Limestone Formation has increased the likelihood of slope failure in several ways. Firstly, it has allowed water to accumulate in open fractures, increasing the hydrostatic pressure and decreasing their stability as well as allowing freeze–thaw processes to operate in severe weather. Secondly, it has promoted the accumulation of clayey weathering products along the discontinuities significantly reducing the friction angle. Thirdly, the presence of cavities is likely to cause irregularities in a cut face, increasing the number of slope orientations on which failures along the major discontinuities can occur. Fourthly, the presence of large pockets of unconsolidated cavity filling can form local unstable areas on a cut face which may collapse or flow outwards, inducing collapse of the overlying rocks previously supported by the infilling material.

### SOLUTION EFFECTS IN OTHER DEVONIAN FORMATIONS

The Plymouth Limestone Formation forms the major limestone deposit within the district, but dispersed limestone beds also occur within the Saltash and Torpoint formations. Most of these limestone beds are too thin and dispersed for any solutional activity to have been significant. In the Millbay area of Plymouth [4693 5428], however, massive limestone beds are locally present within the Torpoint Formation. North-west-trending faults cutting this formation, in association with a (now infilled) sea inlet, have permitted solution of these massive limestone beds, to form significant cavities which appear to be interconnected with the sea.

A few limestone beds, up to 2 m in thickness, are present within the Saltash Formation, as in a quarry [3422 5673] near Polbathic. Small phreatic tubes, less than 0.5 m in diameter, are present in the limestone beds in this quarry.

## Superficial deposits

Superficial deposits in the Plymouth city area are mostly alluvial silt and silty clay with some sand and gravel and few peat beds. According to the GAPS (1985) report these deposits have relatively high liquid limits (mean 52) and moderate plastic limits (mean 31) giving a mean plasticity index of 21. They have a relatively low mean bulk density of 1.76 with natural moisture contents which can exceed 60 per cent and are geographically highly variable. The alluvium can therefore be characterised as being generally soft and in places highly compressible.

## COASTAL SLOPE FAILURE

Inland in the district there is generally only a local thin cover of superficial deposits on the Devonian and Carboniferous bedrock. Because most of this superficial cover is present on valley floors and the lower slopes of the valley sides, failure within these inland superficial deposits is both rare and of very small extent. Along the coast, however, substantial thicknesses of coastal head have been piled against Pleistocene cliff lines on top of the raised beach platform. Erosion of the cliffs has removed much of the support for these head accumulations and in places they have consequently failed by landslipping. At Downderry [315 540], Sims and Ternan (1988) reported that where the head deposits were between 6 and 8 m thick, failure was by the development of tensional fractures parallel to the cliff line. These fractures formed a few metres behind the cliffs and slowly widened until a small section of the head collapsed onto the beach. The rate of coastal retreat caused by this failure was measured by them as averaging 107 mm per year between 1960 and 1988. Where the head was up to 25 m thick at the western end of Downderry, failure took place catastrophically by rotational landslipping on a basal plane within a clay layer at the base of the head. A single landslip event in 1974 caused the cliff to retreat 11 m.

In Portnadler Bay, coastal head up to 30 m thick rests against a buried cliff line in the Bovisand Formation on top of a raised beach platform. Two landslips [2415 5173 and 2435 5183], approximately 150 m wide and 100 to 130 m deep, have formed using the buried cliff line as the backplane of the slips. At these locations bedding and cleavage within the Bovisand Formation in the raised beach platform dip seawards at gentle to moderate angles and the basal plane of the landslips lies below the level of the raised beach platform, within the Bovisand Formation itself. This has caused localised rotation of the bedrock involved within the slips.

# THREE

# Concealed geology

## PRE-VARISCAN BASEMENT

In most other parts of England and Wales 'concealed geology' refers to the unexposed lithostratigraphical units within the 'layer-cake' stratigraphy of relatively shallow intra-continental sedimentary basins. The rocks of the Plymouth district, however, constitute a passive margin sequence. Because passive margin sequences form at sites of continental rifting and are accompanied by extension and rapid subsidence of fault-bounded basins (see Jackson and McKenzie, 1983; Gibbs, 1984) they do not contain a layer-cake stratigraphy. Instead they are characterised by mud-dominated lithologies of extreme thickness, and display considerable thickness variations and rapid facies changes, with fundamental differences in stratigraphy between adjacent fault-bounded basins. These characteristics of the lithostratigraphical units within the district mean that the 'concealed geology' of the district mostly concerns the rocks which form the basement to the passive margin basins.

### Regional tectonic evidence

The thicknesses of the sedimentary fill in passive margin basins commonly exceeds 4 km (see Evans, 1990) and so basement rocks are not expected to be within the range of even the deepest boreholes within the district. Moreover, the deformation of this passive margin by continental collision in the Variscan orogeny has caused sequences to be overthrust, with the underlying basement buried still deeper (see Chapter 11). The nature of the basement rocks can, however, be inferred both from an interpretation of the regional tectonics of south-west England and northern France, and using a variety of geophysical techniques.

The rocks of the district form part of the northern passive margin of the oceanic Rhenohercynian Basin, which opened and closed perpendicularly between the earliest Devonian and the early Carboniferous (Holder and Leveridge, 1994). The rocks forming the pre-Devonian basement of the active southern margin of the Rhenohercynian Basin are therefore also likely to underlie the northern passive margin. Pre-Devonian rocks are represented in Brittany, and in south Cornwall within Devonian olistostromes as fragments eroded off the active margin basement during basin closure (Holder and Leveridge, 1986b). These basement rocks comprise a range of schists and granite together with Ordovician quartzite, and Silurian and Lower Devonian limestone representing a shallow marine shelf sequence (Hendriks, 1937; Sadler, 1973).

Precambrian basement beneath the passive margin sediments of south-west England is represented by granitic xenoliths within the Devonian lava and intrusions of west Penwith which display neodymium depleted mantle model ages DPMD of 625 Ma and 851 Ma (Goode et al., 1987), broadly corresponding with the ages of the earlier Cadomian and the Pentevrian basement in Brittany (Durand, 1977).

The similarities between the ages of the Precambrian basement and the lithologies and facies of the pre-Devonian cover suggest that the passive margin rocks of the district may be underlain by basement the same as that presently exposed in Brittany.

### Geophysical evidence

No boreholes currently reach the pre-Variscan basement of the district and so the results of a number of different geophysical techniques, including gravity, magnetic, magneto-telluric sounding and seismic refraction surveys, have been integrated to investigate the nature and depth of the basement.

Gravity data comprise a systematic regional survey by BGS, which has a coverage onshore of one gravity station per two square kilometres, and dense network of shipborne gravity data offshore, both within and adjacent to the district. These data have been published as 1:250 000 scale maps (BGS, 1985a). Local gravity and magnetic data have been collected over the Plymouth Limestone immediately adjacent to the eastern boundary of the district in a survey by Shelton (1987). The Bouguer gravity field is dominated by a general gradient, decreasing to the north-north-west, which results mainly from the low-density granitic batholith exposed to the north of the district (BGS, 1985a). This gradient terminates just to the south of the district, where there is an east–west belt of maxima across Plymouth Bay, apparently offset by faulting to the south of Plymouth. This belt is continuous with anomalies associated with the high density metamorphic rocks of the Start Complex and the gneissic rocks of the reefs around Eddystone.

In the residual gravity anomaly shown in Figure 9, the main features are reduced in amplitude, allowing smaller anomalies to be discriminated more readily. This was produced by removing a regional field, derived from a smoothed version of the observed anomaly. An example of the lower amplitude anomalies is the smooth, continuous 'high' with a 'low' to the north, running west-north-west from Rame Head and which appears to be caused by a body at some depth. It overlies Lower Devonian rocks, which have a lower density than Middle and Upper Devonian rocks, indicating that there is a higher density body beneath. Elsewhere, in the north of the district a local residual minimum of less than 2 mGal appears to be associated with the Carboniferous rocks of the

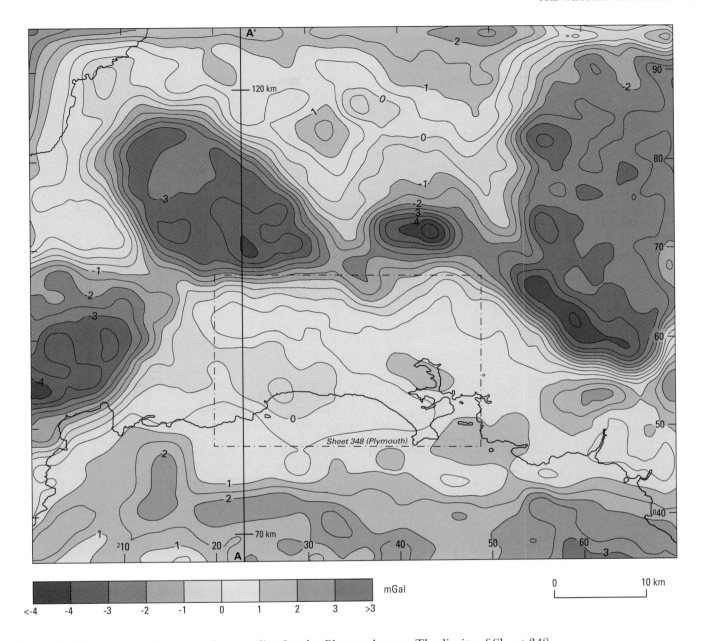

mGal

0          10 km

**Figure 9** Bouguer gravity residual anomalies for the Plymouth area. The limits of Sheet 348 Plymouth are shown as dashed lines. The line of the cross-section section in Figure 12 is shown as A-A'; distances along the section are indicated.

St Mellion area which are tentatively suggested to be less than 700 m in thickness.

Aeromagnetic data, derived from BGS onshore and offshore analogue surveys flown in 1958 and 1961, have been published as 1:250 000 scale maps (BGS, 1985b) and digitised by capturing the position of the intersection of the flight lines with contours (Smith and Royles, 1989). The magnetic field is fairly smooth across the district and decreases slightly to the south (see BGS, 1985b). The residual field for the district (Figure 10), derived by removing a regional field based on a smoothed version of the observed data, displays more

detailed variation, although north–south features are probably due to imperfect adjustment of the data.

To the north of the district is a belt of high-amplitude anomalies with steep gradients which border the northern margin of the Bodmin and Dartmoor granites. At Merrymeet, approximately 5 km north-east of Liskeard, a sharp, linear anomaly is described in more detail below. A series of weak, discontinuous anomalies, which fall along an east-south-east trend [between 200 660 and 400 580] (Figure 10) appear to be caused by near-surface bodies, probably doleritic intrusions and lavas. In the north-east corner of the district, a weak magnetic feature, possibly

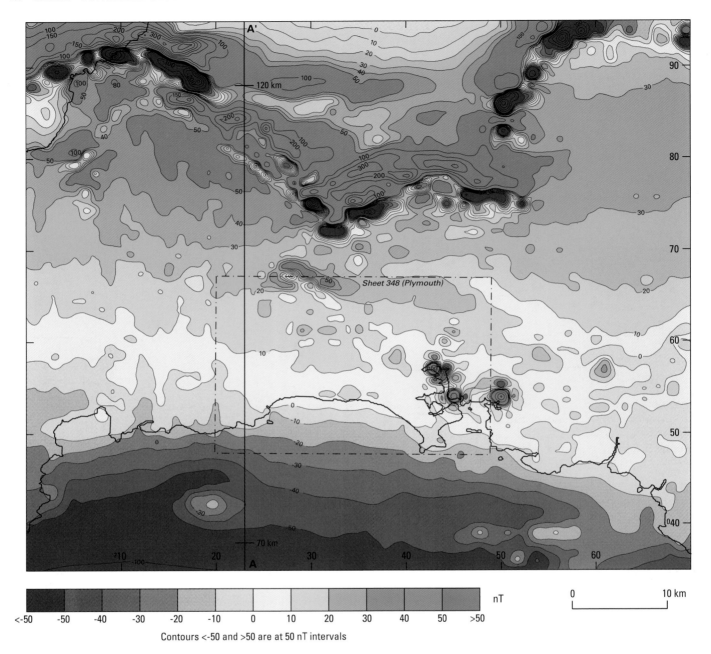

**Figure 10**   Magnetic Total Field residual anomalies for the area. The line of the cross-section in Figure 12 is shown as A-A'; distances along the section are indicated.

caused by a body at some considerable depth, cannot be correlated with any known geological feature and may be related to the buried southern boundary of the Dartmoor Granite. Along the southern margin of the district there is a steepening of the magnetic gradient across a north-of-west-trending line for almost the entire width of Plymouth Bay. The feature is smooth and persistent suggesting an origin at moderate depth.

A number of very strong, steep-sided anomalies are present within, and close to, the urban area of Plymouth.

Their characteristics show them to be laterally impersistent and to be at, or close to, the surface and are therefore most likely to be caused by the concentration of steel associated with the dockyard, the (now disused) power station in eastern Plymouth and with a National Grid distribution station on the north-west of urban Plymouth.

Magneto-telluric sounding data have been gathered from a north–south traverse between the Bodmin Granite and Looe (Beamish and Smith, 1994). This

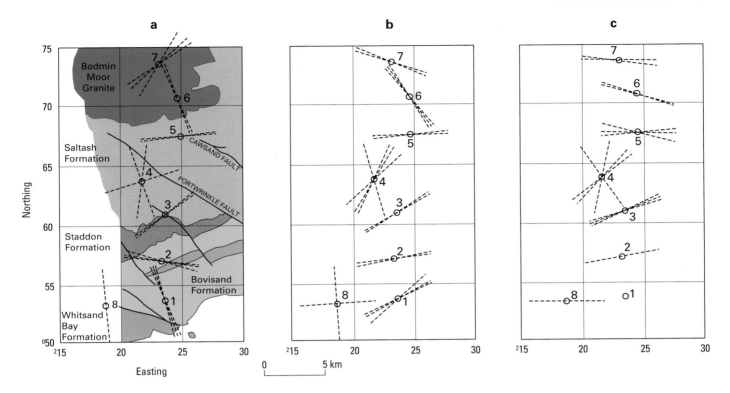

**Figure 11**   Principal geoelectric anisotropy strike orientations from a magneto-telluric sounding survey on the western side of the district.

a. High frequency data for the interval 130 to 78 Hz. This data represents geoelectric anisotropy in the upper 1 km of the ground profile.
b. High frequency data for the interval 22 to 7 Hz. This data represents the upper 3 km of the ground profile.
c. Low frequency data for the interval 0.02 to 0.01 Hz. This data is a representation of the 'whole crustal' geoelectric anisotropy. The variability of orientations across the range of frequencies at site 4 may represent a lack of anisotropy or a null point in the profile.

traverse approximately followed Line 5 of the South West England Seismic Experiment through the district (Brooks et al., 1984). The results (summarised in Figure 11) indicate that the predominantly east–west strike of bedding, thrust strike and cleavage reflects a predominantly east–west structural grain below 8 km in depth, within what is assumed to be pre-Variscan basement.

In order to draw these data together, a cross-section (Figure 12) has been constructed to show how a geological model might satisfy the observed gravity and magnetic data, constrained by magneto-telluric and seismic results. Physical properties of the rocks in the district are shown in Table 8. There is uncertainty in some of these values, which taken together with the complexity of the geometry of the sedimentary basins and the later repeated fold and thrust deformations during the Variscan orogeny, present some difficulty in defining the geophysical model.

The overall thickness of the crust is estimated to be approximately 28 km, assuming generally accepted lower crustal and mantle densities.

The shape of the granite batholith, as modelled from the gravity data, shows a steep southerly dipping southern boundary to the Bodmin Granite extending down to at least 10 km depth, beneath the northern part of the district, as also shown by Rollin (1988, fig. 4). This interpretation differs from that of Bott and Scott (1964) and Al-Rawi (1980), who suggested the presence of a horizontal shelf of granite at a depth of 4 km beneath the northern part of the district. The west-north-west-trending gravity feature, attributed by these authors to this granite shelf, is shown on the Bouguer gravity residual plot (Figure 9) as extending well away from the granite and appears to relate to relatively high-density Devonian sedimentary rocks identified as 'Devonian 2' in Figure 12. This feature is 3 km wide and is smooth and continuous, indicating that the density contrast extends to depth. It appears to coincide with the crop of the slaty mudstone and volcanic rocks of the Saltash and Torpoint formations. In the gravity and magnetic model (Figure 12) the 'Devonian 2' body is shown as being partly overlain on its southern flank by a 'Devonian 1' unit. The crop of

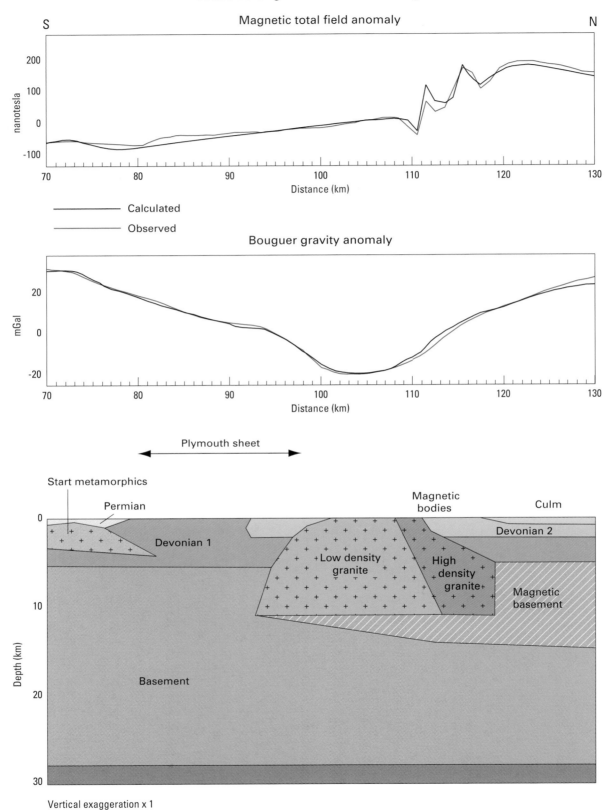

**Figure 12** A crustal model of the gravity and magnetic data along a north–south section through the district. See Figures 9 and 10 for location.

this unit corresponds to that of the Lower and Middle Devonian Saltash, Bovisand and Whitsand Bay formations which are overthrust northwards over the Torpoint and Saltash formations. The presence of the 'Devonian 1' unit beneath the 'Devonian 2' reflects a simple model and should not be taken as an indication that the Lower Devonian formations necessarily underlie Middle and Upper Devonian rocks in the area south of the Bodmin Granite.

The parallel gravity and magnetic features along the southern margin of the district have been referred to above. The model (Figure 12) suggests that they are related to metamorphic rocks of similar type to those in the Start Complex and at Eddystone, which continue northwards for several kilometres beneath the Devonian sedimentary rocks. This interpretation of the data is consistent with the interpretations of gravity and seismic data by Bott and Scott (1964) and Doody and Brooks (1986). Interpretation of detailed gravity data in the Start area suggests that the northern boundary of the higher density Start Complex rocks dips northwards at 45° and extends to a depth of 3 km.

A major long-wavelength magnetic anomaly occurs to the north of the granite batholith. This anomaly is interpreted as being caused by a normally magnetised body below a depth of 5 km, constrained to the south by its contact with the granite (Figure 12). The body appears to form a floor to the granite, wedging out to the south beneath it, at a depth of approximately 12 km. Although this body cannot be related to a particular geological feature the overall structure appears to indicate the presence of structures within what is assumed to be pre-Variscan basement at 12 km depth.

## OTHER CONCEALED FEATURES

North and east of Merrymeet [280 660] a 7 km-long east-south-east-trending magnetic anomaly, with a maximum amplitude of 150 nT, lies parallel to the Ludcott Fault (Figure 13). The anomaly lies within the crop of the Upper Devonian to Lower Carboniferous Burraton Formation and is underlain at relatively shallow depth by the Bodmin Granite (Rollin, 1988). At its north-west end the anomaly is truncated at the contact of the granite. Two-dimensional modelling of the feature suggests that it may be caused by a steep-sided body between 200 and 300 m in width, extending from the granite roof to within 100 m of the ground surface. The bodies exhibit a strong remanent magnetisation, the orientation of which is similar to that measured for Permian rhyolitic dykes and lava at Withnoe and Cawsand (Cornwell, 1967a; BGS, 1985c), Permian lava near Exeter (Piper, 1988), and pyrrhotite mineralisation in basic sills and sedimentary rocks of Carboniferous age near Tavistock (Cornwell, 1967b). Although there is no unambiguous evidence for the nature of this body, the most likely cause is metasomatic pyrrhotite mineralisation, associated with the emplacement of the granite.

**Table 8** Assumed values of rock properties from Bott et al. (1958), Al-Rawi (1980), Shelton (1987), Rollin (1988) and BGS surveys.

| Rock units (see Figure 12) | Density | Susceptibility (SI units) | Remanent magnetisation |
|---|---|---|---|
| Permo-Triassic sediments | 2.55 | | |
| Culm | 2.67 | | |
| 'Devonian 2' (Middle and Upper Devonian) | 2.76 | | |
| Plymouth Limestone | 2.72 | | |
| 'Devonian 1' (Lower Devonian) | 2.68 | | |
| Start Hornblende-schist | 2.85 | 0.002 | |
| Start Mica-schist | 2.70 | 0.002 | |
| Granite (southern part of Bodmin) | 2.65 | | |
| Granite (northern part of Bodmin) | 2.68 | | |
| High-level magnetic body north of the Bodmin Granite | 2.76 | 0.02 | A/m         0.8 Inclination  220° Declination  -7° |
| Deep basement north of the Bodmin Granite | 2.85 | 0.03 | |

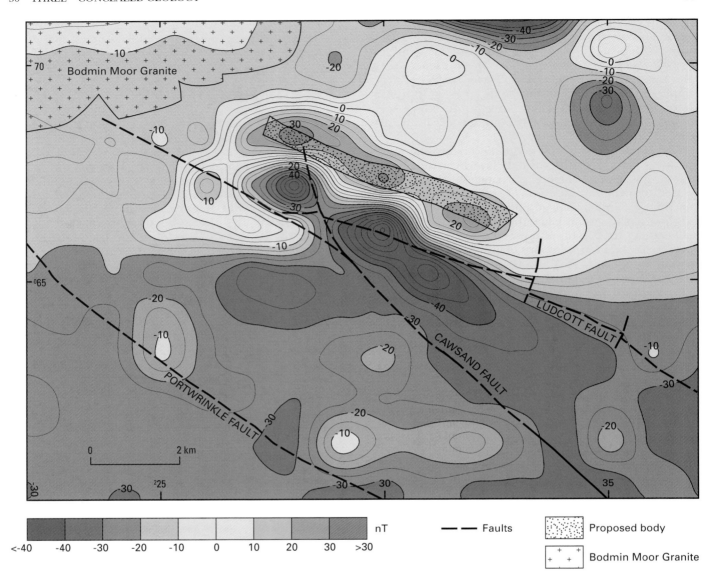

**Figure 13**   Magnetic anomaly map of the Merrymeet area. Contours are in nT. The subcrop (stipple) of a proposed steeply dipping tabular magnetic body is shown. This is probably a zone of sulphide or magnetite mineralisation.

# FOUR

# Devonian

The Devonian strata of this district are recognised in the Looe, South Devon and Tavy basins, and on the Plymouth High (Table 1). The continuity of succession in any of the sequences is not certain due to the general paucity of faunas, nature of exposure and structural complexity.

The major divisions of the Lower Devonian in the region, identified by Ussher (1890) were the Dartmouth Slates, Meadfoot Beds and the Staddon Grits. In the Plymouth district he also recognised (1907) the 'Plymouth limestone', to the north of the crop of the Staddon Grits. He assigned other rocks to the Middle or Upper Devonian on a chronostratigraphical basis.

The Dartmouth Slates have since been elevated to group status and their component formations in south Devon described (see Dineley, 1966; Hobson, 1976; Seago and Chapman, 1988, Smith and Humphreys, 1991). The Meadfoot Beds and Staddon Grits are now recognised as formations of the Meadfoot Group (Harwood, 1976). The Meadfoot Beds of the Plymouth district have been termed the Bovisand Beds (Harwood, 1976) and more recently the Bovisand Formation (Seago and Chapman, 1988); the Staddon Grits are here termed the Staddon Formation. The two groups reflect the major environmental transition from the essentially continental facies of the Dartmouth Group to the largely marine Meadfoot Group.

The Tamar Group is a new grouping of formations and this occurs to the north of the overthrust Staddon Formation. It includes the Plymouth Limestone Formation but mainly comprises the predominating slaty silty mudstones of the Saltash, Torpoint, Tavy and Burraton formations. The established age range of the group is between late Early Devonian (late Emsian) and Late Devonian (late Famennian).

## DARTMOUTH GROUP

In this district, the two components of this group are the Whitsand Bay Formation and the Bin Down Formation. In the coast-exposures, the former contains distinctive subfacies but these cannot be mapped separately inland. These two formations are, respectively, suites of predominantly continental clastic sediments, and turbidites with hemipelagic mudstones and volcanic rocks.

These rocks form the hanging walls of two thrusts which bound tracts to the west of the Cawsand Fault. These strike belts are cut and dextrally displaced by the Portnadler Fault and the Portwrinkle Fault together with its splay, the Rame Fault. The northern crop thins out against the bounding thrust towards the western margin of the district. The wider southern crop is exposed sporadically in the coastal sections west of Portnadler Bay

to the Rame peninsula. The Whitsand Bay Formation constitutes the major part of the group whereas the Bin Down Formation, which occurs in the northern tract west of the Portwrinkle Fault, is present as an intercalation towards the top of the Whitsand Bay Formation.

### Whitsand Bay Formation

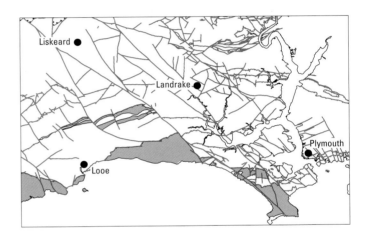

The formation is well exposed along the coast within the southern thrust sheet but the base is not seen. In these sections the sequence is generally the right way up, younging and dipping gently south-eastwards. The lower part of the southern thrust sheet is well displayed south of Looe. In this thrust sheet, the formation is exposed between Millendreath and Portwrinkle and it presents a dip section along the coast of Whitsand Bay between Freathy and Rame Head, although faulting and folding cause the top of the formation to be present at both ends of the section. The upper transitional boundary with the overlying Bovisand Formation of the Meadfoot Group is exposed in the extreme south-west of the district, in Tregantle Cliffs, and on either side of the Rame headland. The maximum thickness of the formation is some 3300 m.

Within some 500 m of the bounding thrust, beds are inverted by a large-scale primary fold verging and facing north-westwards. Secondary folding causes overturning of the central and upper parts of the thrust slice in the poorly accessible coastal section about Polperro. Here, beds dip northwards and face southwards.

To the east of Portwrinkle, the northern tract of the formation is exposed in the coastal section between the Portwrinkle Fault and the Rame Fault splay. A faulted inlier of the formation occurs between Portwrinkle and Crafthole, and an antiformal core flanked by Bovisand

Formation is present between Oldhouse Cove [367 536] and Black Ball Cliff [371 536].

## LITHOLOGIES

Five lithological components have been recognised: reddish purple mudstone and silty mudstone, green silty mudstone to muddy siltstone, green to mauve siltstone, grey-green with some purple sandstone, and off-white to pale green quartzite (Jones, 1992; Barton et al., 1993). There are also subordinate grey silty mudstone beds, locally with thin quartzose sandstone interbeds, that are more numerous towards the top of the formation.

The purple mudstone forms units up to 25 m thick but commonly beds are less than 2 m thick. Purple mudstone beds locally contain pale green reduction spotting. Many beds show other colour mottling or a more discrete 'blebby' character which is imparted by compositional variation. Some of these beds are 'pebbly' mudstones with dispersed small clasts of siltstone and fine sandstone (up to about 50 mm long in the tectonic extension direction). Others contain intensely folded and disrupted, siltstone and sandstone laminae, and yet others are bioturbated, with siltstone and sandstone filling burrows and tubes. This lithology is typically interbedded with siltstone and sheet sandstone.

The green silty mudstone and muddy siltstone, varying in colour from pale yellowish green to blue-green, locally shows a purple mottling. Bedding units of 5 to 10 m thick are common and exceptionally units up to 30 m thick are present. Very rare sand-filled desiccation cracks are present in this lithology, which is also locally 'blebby'. This is commonly due to the presence of darker green sandstone as wisps, disrupted laminae or burrow fills. In other instances it is due to sharply defined chloritic inclusions that locally are associated with a volcaniclastic component, for example south-east of Wiggle Cliff [at 4184 5006].

The pale grey-green to mauve siltstone forms massive beds up to 6 m thick generally with little internal structure, but where coarse siltstone and fine sandstone wisps, lenses and laminae are present, abundant burrows are locally apparent. The last are simple, unbranched, bedding-planar to near-vertical, exichnial and hypichnial burrows (Martinsson, 1970), filled with coarser sediment. Coarse siltstone and fine-grained sandstone are present as discrete interbeds with sharp bases and tops; local sedimentary structures include parallel lamination, ripple cross-lamination and climbing ripple cross-lamination. Organic debris, fish fragments and black phosphatic material are sporadically abundant. Generally the siltstone is interbedded with sandstone or contains fine laminae, wisps or lenses of sandstone.

The sandstone is predominantly grey-green but may locally be pink or purple in colour. Fine-grained greywacke predominates but more quartzose sandstone also occurs. These lithologies form beds, up to 0.8 m thick, and composite units up to 4 m thick. The sandstone beds are predominantly fine grained, but are locally conglomeratic above erosional bases, with mudstone clasts predominating; the tops may show delayed grading. The beds are generally sheet-like: thin beds commonly show parallel lamination, and thicker beds exhibit Bouma (1962) a–c divisions and some hummocky lamination. In the thick units, there are coarser sandstone laminae and beds with downcutting bases, forming load casts in the finer sandstone below. Such units generally display low-angle cross-bedding, climbing dune bedding and channelling.

The off-white to pale green quartzite occurs as isolated beds, up to 0.5 m thick, and within packets of beds up to 4 m thick. Both thinning and fining upwards features are common in these units. Individual beds may be graded and many display low-angle cross-lamination or undulating, bedding-parallel lamination, and tend to be lenticular with erosional bases. Bedding planes commonly form the boundaries to single sets of planar cross-bedding. The lateral passage of one or more impersistent beds to a thicker packet of beds with a downcutting channelled base is common, for example on Looe Island [2556 5156]. The measured palaeocurrents (Jones, 1992) in these sandstones are dominantly unidirectional with vector means that vary from north-west to south-west.

These lithologies occur within three broad component groupings of the formation. In its upper and lower parts, purple and green, finer grained rocks predominate but the central part is characterised by siltstone and sandstone. Locally within these broad divisions some cyclicity is apparent, particularly towards the top of the sequence both in the northern tract at Portwrinkle and in the southern belt, in Hoodney Bay and Polperro, where there is interbedding of grey silty mudstone and fine sandstone.

Towards the top of the formation, a cyclicity is locally well developed, as in the cliffs of Finneygook Beach [360 538], in the lower thrust sheet on the eastern side of the Portwrinkle Fault. Here sequences, a few metres thick, of grey silty mudstone, green silty mudstone, mottled purple and green silty mudstone, and green silty mudstone, with thin quartzose sandstone beds in the grey and green rocks, are repeated over several tens of metres. A similar cyclicity, observed at localities to the west of Looe, is illustrated in Figure 14.

## LITHOFACIES ASSOCIATIONS AND DEPOSITIONAL ENVIRONMENT

Dineley (1966) considered that the rocks of the Dartmouth Group represented stable coastal mud-flat, fluviatile and lagoonal deposition, but studies in the Plymouth district and in coastal sections to the east of The Sound have revealed a more dynamic depositional setting (Smith and Humphreys, 1989, 1991; Jones, 1992). These authors described a setting of perennial lakes in rapidly subsiding basins, with mass flows generated by collapse of unstable basin margins; distal fluvial regimes were also recognised. The lithofacies associations typical of lacustrine, distal fluvial, sheet-bedded and channel regimes are recognised throughout the formation and were described in detail by Jones (1992). The mudstone, silty mudstone and massive siltstone with sporadic intercalations of coarser material constitute the lacustrine facies. These sediments represent the subaqueous deposition of clay and silt from suspension with periodic fine-grained turbid underflows. The bioturbation indicates that, at times, the

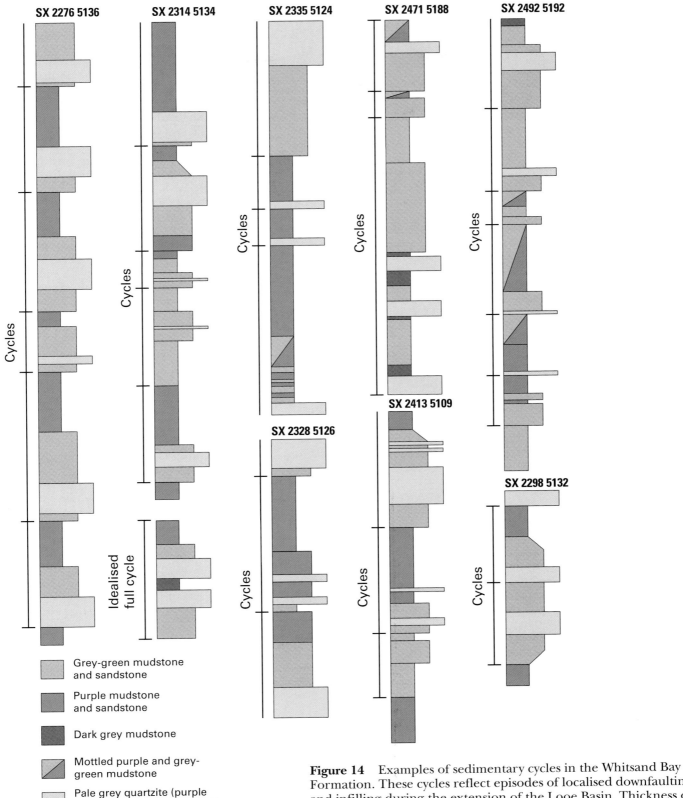

SX 2276 5136

SX 2314 5134

SX 2335 5124

SX 2471 5188

SX 2492 5192

SX 2413 5109

SX 2328 5126

SX 2298 5132

Cycles

Idealised full cycle

Grey-green mudstone and sandstone

Purple mudstone and sandstone

Dark grey mudstone

Mottled purple and grey-green mudstone

Pale grey quartzite (purple stained where it occurs within purple mudstone)

**Figure 14**    Examples of sedimentary cycles in the Whitsand Bay Formation. These cycles reflect episodes of localised downfaulting and infilling during the extension of the Looe Basin. Thickness of the cycles is dependent upon the amount of downthrow and the frequency of downfaulting events, with the full cycle only developed where the maximum downfaulting occurred.

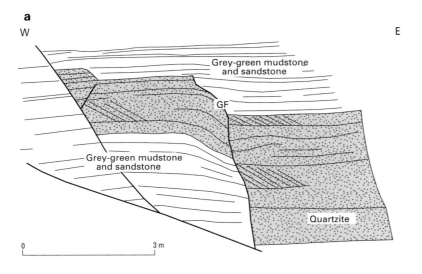

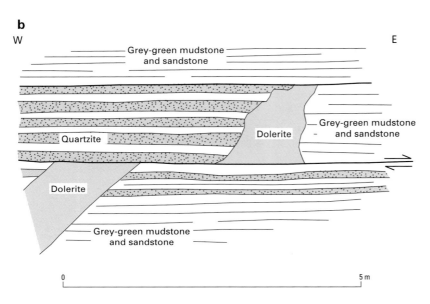

**Figure 15**   Growth-faulting in the Whitsand Bay Formation.

a.   South of Polperro Harbour [2101 5076]. Abrupt change in thickness of a quartzite packet across a growth fault (GF) which is truncated upwards by beds of grey-green sandstone and mudstone.

b.   South-east of Polperro Harbour [2133 5080]. Dolerite dyke intruded along a growth fault laterally truncating a quartzite packet. The dyke has been displaced along a later strike fault, parallel to bedding.

stones probably indicate minor turbid underflows, whereas the thicker beds with gradational boundaries are a product of settlement from suspension following high fluvial discharge. The sedimentary features of the sandstone, such as sharply bounded laminae, lenses and thin beds, reflect emplacement by starved tractional currents and turbid underflows; greater proportions of sandstone within the sequence represent increased fluvial input.

The laterally extensive, sheet-bedded sandstone is interbedded with the lacustrine and distal fluvial facies. The predominant internal sedimentary structure is ripple cross-lamination and the measured palaeocurrents are unidirectional. Amalgamation of beds is common, and thickening-up sequences also occur. These features reflect episodic turbulent underflows with tractional deposition in the lower flow regime.

The features of the quartzite beds, with their erosional bases and large-scale sedimentary structures, result from deposition from confined flows within channels. The presence of unidirectional palaeocurrents and the absence of tidally induced mudstone drapes are consistent with a fluvial environment. The sheet-like form of some isolated quartzite beds or small packets of beds is indicative of channel overflow deposits.

The pebbly mudstone units are not ubiquitous but they are locally numerous in the upper and lower parts of the formation, where finer grained lithologies predominate. These matrix-supported breccias may be up to 8 m thick and contain clasts commonly of a darker hue than the matrix. The dispersion of the clasts indicates emplacement by mass flow. The components of these deposits are not exotic and appear to be the product of collapse caused by intrabasinal instability.

Further evidence of penecontemporaneous instability within the basin is presented at several localities. Some faults at high angles to bedding, and locally with displacements of a few metres, are truncated upwards by bedding surfaces. These are growth faults which, in some instances, bound small contemporary graben filled by fluvial quartzose sands (Figure 15). In this same area, about Looe and Polperro, there are also numerous examples of dolerite emplacement into unconsolidated

water/sediment interface was oxygenated, and the fish remains show that the body of water supported nekton. Simple unbranched burrows are well documented within nonmarine sediments and intertidal environments (Seilacher, 1967; Eagar et al., 1985).

The siltstone and sandstone units, of varying relative abundance, form a lithofacies association of the distal fluvial regime. These lithologies represent an increasing fluvial influence and an input of coarser sediment into a regime of siltation, in which there was abundant bioturbation. The thin intercalations of coarse silt within the silt-

sediments, where irregular apophyses of dolerite extend into locally destructured sedimentary rock. A genetic association appears probable.

Considering the sequence as a whole, the general paucity of features characteristic of emergence such as desiccation cracks, abundant mudstone clasts generated by desiccation break-up, palaeosols and evaporitic minerals, indicates that deposition was in a perennial aqueous environment. Subaqueous deposition is indicated by the various lithofacies described, apart from the channel facies, which could be subaqueous or subaerial. Marine faunas thought by Evans (1981) to be from the formation are from rocks now mapped within the Meadfoot Group. The absence of sedimentary facies typical of the marine environment, such as wave-, tide- or storm-generated sequences, but the presence of dispersed nonmarine bivalves and gastropods (Ussher, 1907; Dineley, 1966; see below) point to a predominantly freshwater regime, a perennial lake setting. Such thick and extensive sequences of lacustrine mudstone, with sparse evidence of emergence, are indicative of large, continuously subsiding basins in a tectonically active area. The sedimentation was directly related to the influx of fluvial sediment which was supplied principally from the east. It is possible that the lake was hydrologically open with a marine connection, particularly in later times, when grey mudstones were deposited within the cyclic units, which were probably tectonically influenced.

Smith and Humphreys (1989, 1991) have described similar lithologies and facies in the Dartmouth Group immediately to the east of this district and drew similar conclusions. They did, however, note the presence of subaerial sheet-flood sandstone. Also the proportions of subaqueous debris flows and finely heterolithic sequences are greater there. The former carry much coarser clasts, a function of proximality, and the latter indicate mild wave agitation and sedimentation above wave base. All are consistent with a shallowing of the depositional basin(s) to the east of the district.

## Bin Down Formation

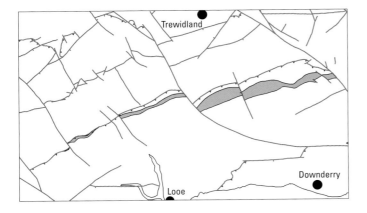

This formation has been mapped near Bin Down [276 578], within the northern thrust belt containing the Dartmouth Group, to the west of the Portwrinkle Fault. Despite poor exposure generally, and non-exposure of boundaries with the Whitsand Bay Formation, it is thought to interdigitate with that formation. Although in contrast it comprises grey sediments, predominantly mudstones, there are several reasons for placing it within the Dartmouth Group rather than within the succeeding group. A borehole sunk near Widegates (SX 25 NE 1) [2748 5747] has shown its upper boundary to be transitional in a slaty mudstone, siltstone and sandstone sequence. Some observations of relationships between bedding structures and early ($D_1$) cleavage near the base of the formation [for example at 2679 5746] show that the sequence is the right way up, with no significant overturning.

The formation is distinctive because it contains a significant suite of extrusive volcanic rocks, probably more extensive than has been delineated on Sheet 348, Plymouth. These rocks include basic vesicular lava, massive to thinly bedded hyaloclastite and tuff. Tuff and crystal tuffite of more acidic aspect have also been mapped, but only from surface rock fragments.

On Bin Down, the formation is about 300 m thick, but thins both to the east and west. Westwards, the main crop terminates against a north-west-trending fault near Tredinnick [238 571] but reappears above the basal thrust to the south of Muchlarnick [between 230 567 and 218 561]. Eastwards, the crop is terminated by the Portwrinkle Fault. On the eastern side of this fault, the northern belt of Dartmouth Group rocks is represented to the east and south-east of Portwrinkle, where two antiformal cores of the Whitsand Bay Formation are separated by a synformal core of Meadfoot Group rocks. Here, the northerly inlier of the Whitsand Bay Formation is thought to lie at a similar stratigraphical level to that which includes the Bin Down Formation, to the west of the Portwrinkle Fault, but it differs in structural level.

### LITHOLOGIES

The formation comprises grey to dark grey, silty mudstone, which is variably interlaminated or interbedded with siltstone; these are interbedded with pale grey quartzose sandstone, of variable bed thickness. Locally, there are thicker sandstone sequences comprising amalgamated beds, as at Tregarland Tor quarry [2522 5754]. In the Widegates Borehole, which penetrates the top half of the formation where it is at its thickest, grey slaty mudstone with interbedded sandstone are present above and below a clastic volcanic sequence some 40 m thick. Phosphatic nodules are locally common in the mudstone. The sandstone forms thin to medium beds, commonly showing normal grading, scour structures, parallel and cross-lamination, and rip-up clasts in coarse-grained basal units. Some adjacent beds show multiple graded lamination, from coarse-grained sandstone to silty mudstone, others are wavy laminated (Ksiazkiewicz, 1954). Load casting, dewatering structures, slump folding and dislocation of bedding are common. In the lowest part of the borehole sequence, there are thin beds of sedimentary breccia, with clasts of sandstone, siltstone and mudstone, interbedded with sandstone.

Ussher (1907, pp.20, 26) reported that 'hard grits' on Bin Down and Corgorlan (Tregarland) Tor yielded fish remains, much the same as those of the Whitsand Bay Formation.

The predominantly basic, extrusive volcanic rocks, form a locally very variable proportion of the formation. The sequence in the Widegates Borehole contains massive, coarse-grained hyaloclastite, thin- to medium-bedded hyaloclastite, lithic tuff and tuffite. The basaltic rocks are locally highly silicified and are associated with sulphides.

DEPOSITIONAL ENVIRONMENT

The sedimentary structures of the sandstone beds are characteristic of turbidite emplacement, and the general instability of the depositional regime is indicated by the slump structures and sedimentary breccia. The hyaloclastite indicates a subaqueous environment, whereas the phosphatic concretions in the mudstone point to a marine setting (Humphreys and Smith, 1989). The volcanic activity may be linked with instability within the depositional basin but the apparently abrupt transition of facies and marked local thickness variations suggest the possibility of growth faulting, causing a marine incursion. It is possible that the thick, fish-bearing sandstone sequence of Tregarland Tor may represent a fluvial input and build-up in a shallow margin of the basin.

This formation is similar to the Yealm Formation (Dineley, 1966) of the Dartmouth Group, to the southeast of Plymouth. That formation comprises interbedded quartzose and greywacke sandstone and grey, slaty mudstone with generally up to 20 per cent of volcanic rocks (Dineley, 1966), of both acid and basic composition (BGS, 1990). The proportion of volcanic rocks is considerably greater around Whympston where an acid/basic volcanic complex has been recognised in a series of BGS Mineral Reconnaisance Programme Boreholes (Unpublished BGS data). *Pteraspid* remains are common in the sandstone of the Yealm Formation (Dineley, 1966). The depositional setting of the formation has been interpreted by Smith and Humphreys (1991) as one of mass-flow-dominated lakes with fluvial inputs.

**Biostratigraphy of the Dartmouth Group**

The age ranges of the formations of the Dartmouth Group are not known with any precision. Non-vertebrate faunas are generally lacking, although Ussher (1907) detailed his own and earlier finds of the gastropod *Bellerophon bisulcata*. The group was assigned by House and Selwood (1964) and House et al. (1977) to the Siegenian (Pragian), with the possibility that it extends down into the Gedinnian (Lochkovian), on the basis of the reassessment by White (1956) of early finds of fish remains. He concluded that the *'Pteraspis cornubica'* recorded in Ussher (1907) represented *Protaspis*, *Rhinopteraspis leachi* and *Rhinopteraspis dunensis* that are indicative of the early and mid Siegenian. It transpires from recent mapping that many of the localities from which these fish remains were collected, as recorded by Ussher (1907, p.6), lie within the Bovisand Formation of

the Meadfoot Group. However, recently collected specimens of pteraspid heterostracans from sandstones in the top half of the Whitsand Bay Formation, just north-west of the Long Stone [337 557], have also been ascribed to *Rhinopteraspis dunensis* or *Althaspis leachi* and assigned by Forey (in Ivimey-Cook, 1992) to the early Siegenian.

**MEADFOOT GROUP**

The Meadfoot Beds were defined by Champernowne (1881) as 'bluish grey slates with hard grits' in a small fault-bounded block on the Hope's Nose promontory at Torquay. These fossiliferous rocks were correlated by Ussher (1890) with similar Lower Devonian rocks present elsewhere in south Devon. In the first Plymouth memoir (Ussher, 1907) this subdivision was termed the Meadfoot Group and included some thick sandstone sequences at Looe, namely the Looe Grits. The thicker Staddon Grits were not included in the group but placed above it. Eastwards from The Sound however, this distinction could not be maintained with any certainty by Ussher (1912): all sandstone sequences, some possibly within the Meadfoot Group, were assigned to the Staddon Grits. This led Harwood (1976) to propose the inclusion of the Staddon Grits within the Meadfoot Group, and the redesignation of the 'Meadfoot Beds' as the Bovisand Beds, which are adjacent to the type section of the Staddon Grits on the east side of The Sound. Minor nomenclatural amendments have since been made (Chandler and McCall, 1985; Seago and Chapman, 1988) and these subdivisions are now referred to as the Bovisand Formation and the Staddon Formation.

**Bovisand Formation**

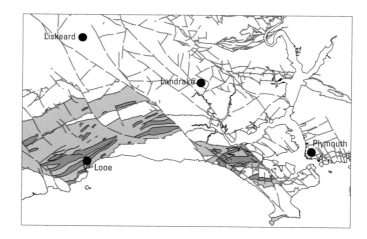

The type section of the Bovisand Formation (Harwood, 1976; Seago and Chapman, 1988) lies just to the south of Staddon Point in Bovisand Bay, immediately to the east of this district. The lowest part of the formation seen hereabouts is strike-faulted against the Dartmouth Group in Crownhill Bay [492 497], and the highest beds are also faulted against the Staddon Formation in Bovisand Bay. The age range of rocks along this coastal section, through the hinge zone and part of the steep northern limb of a major antiformal fold, is not clearly defined. Evans (1983) thought that brachiopod faunas showed the Pragian (Siegenian)–Emsian transition to occur at the boundary of the Bovisand ('Meadfoot') and Staddon formations, whereas Dean (1989a) placed much of the section within the Emsian on the basis of its palynomorph assemblages.

In this district, the formation is present to the west of the Cawsand Fault within three thrust and fault-bounded nappe sequences (Figure 1). The cumulative thickness of the formation is at least 1300 m. The northernmost and central nappe sequences are present in a broad tract, up to 7 km wide, west of the Portwrinkle Fault and they form much of the coastal section between Portwrinkle and Tregantle Cliff, between the offshore extension of the Portwrinkle Fault and the Rame Fault. They also extend eastwards near Millbrook to the east of the Rame Fault.

The southernmost sequence occurs on the southern part of the Rame peninsula where there is a transitional boundary [4180 4170] with the Whitsand Bay Formation. The lowest part of the formation, some 300 m thick, is separated by strike-slip faulting from a differing part, about 400 m thick, forming Rame Head.

Despite the repetition of the formation in these thrust nappes, poor inland exposure precludes a full appreciation of the facies variations and hence of the depositional basin over an extended area. Nevertheless, there is good representation in the cliffs between Portwrinkle and Tregantle, and the cliffs at Long Sands, in particular, provide a section more extensive and complete than the type section.

The formation broadly comprises grey slaty mudstone and siltstone with generally subordinate sandstone, but two major sandstone-dominated sequences constitute the Long Sands Sandstone members. Coarse bioclastic limestone is sporadically present as thin lenticular beds or as thicker cross-bedded units in the sandstone members.

### Basal part of the formation

The base and lowest part of the formation of the central nappe are exposed in the coastal section below Black Ball Cliff [3706 5351] and between Oldhouse Cove [367 536] and Harry's Ball Cliff [3606 5674] (Figure 29b), and in the southernmost and upper nappe near Rame Head and at Tregantle Cliff [388 526] (Figure 29a). The sequences near the formational base in Old House Cove and at Tregantle Cliff differ. At the former locality, the uppermost strata of the Whitsand Bay Formation comprise typical interbedded red and green slaty mudstone, with sporadic thin quartzose sandstone beds, and bioturbated, grey, slaty mudstone containing scattered fish debris and black phosphatic nodules. At the latter locality, packets of quartzose sandstone are more prominent and interbedded grey slaty mudstone is sparse.

In Oldhouse Cove, there is an upward passage to grey slaty mudstone with thin interbeds of grey-green siltstone and fine sandstone; dark grey slaty mudstone becomes predominant over some 20 m. This mudstone is commonly bioturbated, with bedding normal and J-shaped sand-filled burrows prominent. Fine-grained siltstone forms laminae within thick cross-laminated beds. Subordinate medium beds of bioturbated, grey siltstone and sporadic thin to medium beds of grey, fine-grained sandstone also occur. The sandstone beds, up to 0.5 m thick, commonly have gradational bases, sharp fining-upwards tops, current ripple cross-lamination, climbing ripples, flow ripples and wave cross-lamination (compare with de Raaf et al., 1977). Higher in the succession, interbedded and interlaminated siltstone and sandstone predominate locally (Jones, 1995). Pale grey quartzite beds, with basal scour structures and low-angle cross-lamination, are sporadically developed throughout the sequence.

The boundary in Tregantle Cliff is more sharply defined: sandstone occurs as laminae, lenticular and continuous thin beds, load casts and scour fills, within bioturbated, grey mudstone immediately above the base. Black phosphatic nodules (up to 30 mm across) occur in the mudstone, with fish debris in their centres. The sandstone shows cross-lamination, with common asymmetrical ripple-form sets, and some climbing ripples. Within a few metres of the base of the formation, thin to thick beds of laminated fine-grained sandstone are common: these grade upwards into siltstone. Thin beds of carbonate-rich siltstone and siltstone, with dispersed brachiopods and turritellids, are sporadically present. Within a few tens of metres of the base, sandstone is locally dominant, with thin to thick beds, up to 1.0 m, showing basal scouring, cross-lamination and normal grading. Sporadically present through the sequence, there are packets (up to 4 m thick) of quartzose sandstone beds, similar to those in the Whitsand Bay Formation.

To the north and east of Rame Head, the base and lower part of the formation show characteristics of those seen both at Oldhouse Cove and Tregantle Cliff. The base is gradational, where the predominantly green silty mudstone of the Whitsand Bay Formation gives way to the grey silty mudstone of the Bovisand Formation over several metres, both having thin interbeds of siltstone and sandstone. In the Bull Cove [412 485] area, some 300 m above the base, there are several beds of silty mudstone, and fine-grained sandstone, with dispersed crinoid and coral debris and lenses of brachiopod shell casts. Stratigraphically above these beds, there occurs the sequence seen at Rame Head headland, on the southwestern side of north-west–south-east dextral strike-slip faulting through Western Gear [4175 4850] and Eastern Gear [4195 4845], This sequence of interbedded grey mudstone, siltstone and sandstone, about 400 m thick, contains sporadic packets of quartzose sandstone. Some of the sandstone beds lower in this sequence are heterolithic, showing fine interlamination of coarse- and fine-grained sand and silt. At the extremity of Rame Head, planar beds, up to 1 m thick, show Bouma (1962) sequences with grading, parallel laminated coarse bases, massive central parts and cross-laminated fine-grained tops. Locally within

**Plate 1** Shark fin spine recovered from the Bovisand Formation of Rame Head. The fossil belongs to the genus *Ctenocanthus* whose range was formerly attributed to the Carboniferous in Europe. Exposed section is 120 mm in length (GS509).

the intervening dark grey mudstones there are lenses of sandstone with abundant fish debris (Plate 1).

### Lower Long Sands Sandstone Member

Higher in the succession, the Lower Long Sands Sandstone Member extends some 750 m east-south-eastwards along the coast from Black Ball Cliff [371 535]. This part of the coast also provides the best exposure of the succeeding strata, including the Upper Long Sands Sandstone Member, in the central nappe (Figure 29b): the problems of correlation of these members are discussed in the section on the latter. The Lower Long Sands Sandstone Member hereabouts is incompletely exposed with a thickness of 320 m. The lower boundary of this member, at Black Ball Cliff, is a low-angle fault dipping gently south-eastwards. Apart from one small-scale open anticline near this boundary, the member is free from folding and dips gently to the east-south-east. Although sandstone is the dominant lithology, it comprises a series of sedimentary cycles which also include mudstone, siltstone and subordinate limestone. Sporadic packets of quartzose sandstone are also present. Alteration colours of buff, ochre, red and purple colouring are prominent in the section, which rarely shows grey siltstone and greenish grey sandstone as fresh lithologies.

Some 25 component main cycles, ranging in thickness from 4 to 32 m, are present. Typically, each cycle coarsens upwards from silty mudstone or siltstone at its base. Sandstone forms sparse, discontinuous and continuous, planar and irregular laminae and also sporadic thin beds near the base; its proportion increases upwards as bed spacing decreases and bed thickness increases. In the upper part of the cycle, sandstone beds (up to 1.5 m thick) amalgamate to form persistent composite units. They show steep and low-angle cross-lamination or bedding-parallel lamination. The grain size shows a

tendency to coarsen upwards through a cycle; individual beds may however show internal fining upwards.

Both fine- and coarse-grained rocks contain abundant shelly fossils locally. There are thin limestone beds and lenses in the low parts of the cycles but the main limestone occurrences are near or at the tops. The limestone is generally pink to red, coarsely bioclastic, and with whole brachiopod shells up to 0.12 m long, as observed on waterworn surfaces. Individual beds are up to 0.3 m thick, and composite beds are up to 3.0 m thick. Low-angle cross-lamination and channelling characterise the thicker units. In thin sections, these rocks are seen to be biosparite packstone [E 67224, E 67225].

### Strata between the Lower and Upper Long Sands Sandstone members

Between the two members, there are some 150 m of predominantly grey slaty mudstone and silty mudstone with minor proportions of sandstone and limestone. The sequence occurs in a major fold couplet of overturned synform and antiform, verging and facing north-westwards (Figure 29b). There are a few minor parasitic folds in the section but it is faulting that disrupts the stratigraphical continuity. The near juxtaposition of structures across the faults indicates no great displacement.

The slaty mudstone and siltstone are dark grey or grey where unaffected by ochre, red and purple staining. Crinoid debris, shell fragments and whole solitary corals are widely scattered in these rocks. Limestone laminae and beds (exceptionally up to 1.5 m thick) with similar fossils occur within these rocks, and in places occur at regular intervals of some 10 m. They also occur in association with coarser sediments, silty sandstone and sandstone, in coarsening upwards cycles. Individual sandstone beds are up to 0.2 m thick but composite units may be up to 2 m in thickness. The bedding surfaces and internal structures are poorly shown in silty sandstones but, in the thicker sandstone beds, fine cross-laminated sets on a 10 mm-scale are apparent.

## Upper Long Sands Sandstone Member

The Upper Long Sands Sandstone Member dips gently eastwards and is exposed over some 180 m of the cliff section at the south-eastern end of Long Sands beach [384 529]. Neither lower nor upper boundary is exposed, but the latter is thought to be faulted. The thickness of the member here is approximately 100 m. The lithologies are similar to those of the lower member but a cyclicity of bedding is not apparent. The lower part of this member is mainly sandstone, variably thinly to thickly bedded and as composite units; it is also fine to coarse grained and friable to indurated The composite units, up to 5 m thick, comprise mainly small-scale climbing ripple sets. This sandstone is locally richly fossiliferous, with bivalves predominating. Siltstone is present only as sparse thin beds and separating smears, in the lower part of the member. Although generally subordinate to sandstone in the upper part, siltstone with sandy siltstone is present, about 20 m thick, in a section east of the cliff promontory [at 384 528]. These beds are bioturbated and contain wisps, lenses, laminae and thin beds of sandstone, with cross-lamination, comparable with parts of the Lower Long Sands Sandstone Member. The succeeding beds are about 20 m of parallel bedded sandstone, in beds up to 2 m thick and rich in fossil debris. Limestone beds (up to 0.5 m) are present throughout the upper part of the member, some showing internal laminar bedding and hummocky cross-stratified tops.

The upper boundary of the Upper Long Sands Sandstone Member on the coast near Tregantle [386 528] is formed by the basal thrust fault of the overlying nappe. Inland, there is generally insufficient structural information to correlate sequences with those of the coastal section with any precision. However, there is sufficient to indicate the presence of larger scale folds in the sequences of the northern nappe, just to the north-east of the Long Sands section across the Rame Fault, and within the same central nappe, west of the Portwrinkle Fault, about Looe. The thick sandstone units there, the Looe Grits of Ussher (1907), are tentatively correlated with the Long Sands members where their relative positions within the Bovisand Formation can be established.

*Sedimentary facies in the Long Sands section of the Bovisand Formation* A number of different sedimentary facies are recognised within the rocks of the Long Sands coastal section. These are summarised with their interpretation in Table 9.

## Strata above the Upper Long Sands Sandstone Member

In the Looe area, sequences of grey slaty mudstone with interbeds of sandstone (up to 1 m thick) are thought, on structural grounds, to overlie the Upper Long Sands Sandstone Member between the Sunrising Estate [259 543] and the vicinity of Waylands Farm [2312 529]. The thickness of this sequence is unknown but appears to be of the order of several tens of metres.

### BIOSTRATIGRAPHY

The fossiliferous sequence at Bull Cove (see above) yielded a brachiopod fauna to Evans (1981) who considered it to represent a probable late Siegenian (Pragian) assemblage. At the time, the published geological map (IGS, 1977) was based on Ussher's work, where these rocks were assigned to the Dartmouth Group, and Evans deduced that the fauna represented a marine incursion within that group. This horizon is now taken to lie some 300 m above the base of the Bovisand Formation.

A *Ctenacanthus* spine (Plate 1) collected from Rame Head is not diagnostic but these grey beds low in the formation and those in the sedimentary cycles, high in the underlying formation, have yielded a significant proportion of the pteraspids recorded by Ussher (1907).

The limestone and sandstone, now assigned to the Long Sands Sandstone members, yielded abundant shelly faunas during the first survey of the district. These fossils are recorded in the descriptive memoir together with the identifications of previous authors (Ussher, 1907; pp.29-40). Macrofossils are poorly preserved in both limestone and sandstone, but casts of brachiopods do occur.

During this resurvey, these rocks have yielded conodonts which indicated to Dean (1994a, b; personal communication, 1996) that the Lower Long Sands Sandstone Member is probably Pragian and that the upper member is within the Pragian to mid Emsian range: a mid Emsian age {*Icriodus bilatericrescens bilatericrescens–Pandorinellina steinhornensis steinhornensis* conodont fauna of Higgins and Austin (1985)} is favoured. It is probable, therefore, that the Pragian/Emsian boundary falls between the two sandstone members.

### DEPOSITIONAL ENVIRONMENTS

A variety of marine environments are reflected in the lithologies of the Bovisand Formation. The basal strata of the formation contain mudstone and massive siltstone, deposited from suspension, whereas the laminated siltstone and sandstone were deposited from tractional flows, as indicated by the cross-lamination. The sandstone was probably emplaced as turbidite flows, the thinner beds having ripple-form sets. The section at Tregantle shows more evidence of periodic wave reworking than that at Old House Cove: current features predominate including wave cross-lamination and combined ripples. The abundance of mudstone at the latter locality indicates quieter water conditions there.

An offshore setting is indicated for these rocks, with sporadic lower shoreface reworking at Tregantle. The basal boundary does not appear to be marked by a transitional sedimentary facies (Jones, 1995). This probably reflects a rapid marine inundation of a region already dominated by perennial lakes with periodic connection to the marine environment (Smith and Humphreys, 1991).

Three facies associations are recognised higher in the formation at Long Sands (Jones, 1995) and these are described below.

**Table 9**  Sedimentary facies recognisable in the Bovisand Formation, Long Sands.

| Facies | Description | Interpretation |
|---|---|---|
| 1. Mudstone–siltstone | Grey to greenish grey, laminated to massive mudstone and siltstone. Rare sandy siltstone and silty sandstone beds and laminae. Some scattered bioclastic debris, occasionally as lags; bioturbated in places. Rare wave ripples and cross-lamination in sandier parts. | Deposition from suspension in quiet, open marine setting — offshore to shelf environment. Rare storm-deposited bioclastic lags and sands. |
| 2. Cross-laminated sandstone | Sheet-like sandstone up to 0.1 m thick; occurs interbedded with most other facies. May have sharp bases and tops, commonly cross-laminated; rarely massive. Bioturbated in places. | Deposition from unidirectional currents. Sands interbedded with facies 1 probably represent storm-generated turbidites in an offshore setting. |
| 3. Cross-bedded sandstone | Cross-bedded fine- and medium-grained sandstone in beds up to 0.3 m thick. Some parallel lamination or low-angle cross-bedding; rare rippled tops. Occurs interbedded with facies 2 and 4 to 8. | Deposition by 2- and 3-D dunes in a shallow marine, shoreface setting. Low-angle cross-bedding formed at high current velocities; rippled tops indicates fairweather reworking. |
| 4. Wave-rippled heterolithic | Lenticular to thinly bedded silty sandstone and sandy siltstone. Wave ripple cross-lamination and combined flow ripples dominant. Forms units a few metres thick. | Lower shoreface deposits; fluctuating energy regimes result in deposition of bedload and suspended sediment load, reworked by oscillatory flows. |
| 5. Bioturbated sandstone | Fine- to medium-grained sandstone and silty sandstone. Structureless, extensively bioturbated by *Chondrites*, *Planolites* and escape traces. Sedimentary structures preserved in places. | Results from the burrowing and destruction of sedimentary lamination. Shallow marine deposit — probably represents modification of other sandy facies. Low sedimentation rates likely. |
| 6. Hummocky cross-stratified sandstone (HCS) | Fine- to medium-grained sandstone; well-bedded with sets up to 0.15 m thick and wavelength up to 1.7 m. Troughs and hummocks preserved, with form concordant foresets. | Combined flows during the waning stage of storms. Deposition in lower shoreface likely. Form concordant laminae indicates vertical accretion processes were dominant. |
| 7. Amalgamated hummocky cross-stratified sandstone | Fine- to medium-grained sandstone. HCS occurs as a series of amalgamated sets, in beds up to 1 m thick. HCS sets up to 0.15 m thick and wavelength up to 2.3 m. Troughs and hummocks preserved, with form concordant foresets. Some thickening of foresets across hummocks. Upwards passage into wave rippled sandstone. | Combined flows during the waning stage of storms. Thickening of foresets across hummocks indicates vertical bedform aggradation. Amalgamation indicates agitated conditions that does not allow mud deposition; upper shoreface deposition likely. |
| 8. Hummocky cross-stratified conglomerate | Clast-supported conglomerate consisting of crinoid debris. Symmetrical form sets with flat bases and rounded, hummocky crests. Wavelength up to 1.9 m and amplitude 0.1 m. Form concordant foresets. Thickening of laminae across hummocks and thinning into troughs. Interbedded with facies 7. | Interpreted as 'coarse grained ripples' (compare Leckie, 1988) formed by storm-induced combined flows. Interbedding with facies 7 indicates similar processes in operation, in an upper shoreface environment. |
| 9. Skeletal wackestone | Wackestone, in beds typically 0.2 m thick, rarely up to 5 m. Sand and fine carbonate matrix. Bioclastic debris common, including crinoids, corals, bryozoans and brachiopods. Stylolitic and recrystallised. Cross-bedded and cross-laminated in places. Interbedded with all other facies. | Marine carbonate accumulation in environments varying from shallow marine to offshore. Bioclastic debris reworked and transported, deposition involves unidirectional currents. |

**Plate 2** Two shoreface sand bodies within the Lower Long Sands Sandstone Member, Kerslake Cliff [375 533], viewed in an easterly direction. Rucksack is 0.5 m high (GS510).

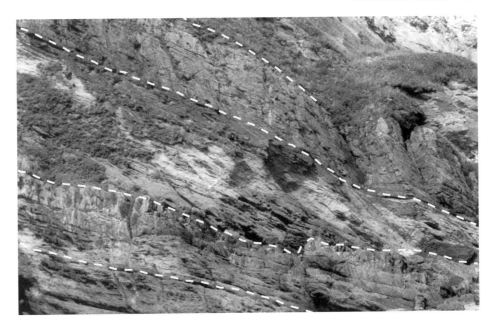

An *offshore-shelf facies* is represented by the mudstone and siltstone, commonly bioturbated and with subordinate limestone and sandstone. The bioclast-rich wacke-stone (facies 9 of Table 9) represents periods of minimal siliciclastic sedimentation permitting carbonate deposition. The presence of sandstone with wave and ripple cross-lamination probably represents shallowing and deposition in the offshore zone.

A *wave-dominated shoreline facies* association (facies 3–6 of Table 9) is represented by the thick sand bodies of the Lower Long Sands Sandstone Member. Interbedded on a large scale with the rocks of the offshore-shelf facies (Plate 2), they form integral parts of the coarsening-upward cycles. Sand was deposited in shallow marine conditions with reworking by shoaling waves. In the absence of progradational delta features, a wave-dominated shoreline setting is indicated. The coarsening-upward successions indicate shoreline progradation. The symmetrical wave ripples are typical of the lower shoreface; the more asymmetrical ripples and dunes form higher up on the shoreface (Clifton, 1976). The introduction of coarse bioclastic debris was probably by storm events responsible for the hummocky cross-stratification.

A *storm-dominated shoreline facies* characterises the Upper Long Sands Sandstone Member. Here the sandstone units, up to 5 m thick, are characterised by wave ripples and hummocky cross-stratification. The hummocky cross-stratification is interpreted as having been generated in storm conditions (Jones, 1995).

Thus the deposits of the formation represent shallow marine deposition, the finer sediments representing offshore shelf environments and the coarser elements indicating more turbulent shorelines. The thick sequences with cyclic developments represent numerous phases of rapid regression and gradual progradation. These features may have been regulated by periodic tectonic activity.

## Staddon Formation

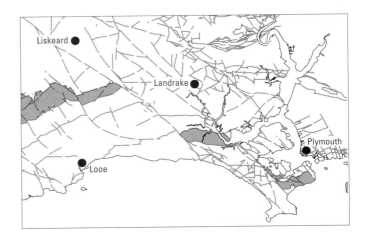

The type section of this sandstone formation is below Staddon Heights flanking The Sound at the eastern margin of the district (see below). The faulted base and top of the formation occur just to the east of the district boundary.

In the western part of this district, the crop is up to 2.5 km wide. Eastwards, the crop of the formation, which commonly forms ridges, is displaced successively by dextral strike-slip faults, mainly the Portwrinkle and Cawsand faults. Both northern and southern boundaries are faulted: the northern boundary is commonly a thrust and the southern boundary a steep normal or reverse fault. The only exposure of a contact is a faulted boundary, moderately inclined southwards near The Bridge in the foreshore [4595 5190] on the west side of The Sound. Here, the north-west-trending strike-slip Mount Edgcumbe Fault intersects an overturned sequence dipping moderately southwards and facing northwards.

The last overturned sequence is at a similar structural level to that at Staddon, which lies within the steep to overturned limb of a major northward-verging fold in which there are parasitic small-scale folds. Elsewhere, all observed sections within the formation, in sporadic quarries and minor cuttings, in the banks of the Lynher River near Erth Hill, at Millbrook and about Cawsand Bay show the sequence to be predominantly the right way up. This indicates that these crops are at a higher structural level than those about The Sound. In general, the lower parts of the formation lie in the northern parts of the crops and the higher parts are to the south.

## LITHOLOGIES

There are two important sections in the district, the Staddon dip-section in Bovisand Bay on the eastern margin, and the other extends from the western side of The Sound into Cawsand Bay. The sedimentology of these sections has been described in detail by Humphreys and Smith (1988). At the first locality, in Bovisand Bay, the lower boundary was mapped [at 492 509] as a fault (BGS, 1996a). A transitional sequence just north of the fault is included within the formation, as observed by Harwood (1976) and Hobson (1978). It comprises grey silty mudstone with thin iron-stained siltstone and thin fine-grained sandstone beds (Bovisand Formation lithologies) interbedded with medium-grained sandstone beds (Staddon Formation lithologies) up to 1 m thick. Medium-grained sandstone becomes dominant in a sequence that thickens and coarsens upwards. The sedimentary structures of the two types of sandstone beds differ; the fine-grained sandstone is characterised by grading, load casting, wave rippling, and faint parallel and undulating lamination. The medium-grained sandstone shows wave ripples with flaser and lenticular bedding, small trough cross-bedded sets, and thick parallel-laminated parts. The transition is accompanied by a decrease in abundance of marine trace fossils, that include *Palaeophycus* and *Rusophycus,* and a disappearance of body fossils (Humphreys and Smith, 1988).

Towards Staddon Point [486 507], the overlying sequence comprises medium- to thick-bedded sandstone with heterolithic interbeds. A fining-up sandstone unit, 20 m thick, comprises a lower 4 m of swaley cross-stratified sandstone with hummocky cross-stratification with penecontemporaneous erosion of the hummocks (Leckie and Walker, 1982); they are succeeded by thick to medium beds of fine-grained sandstone with parallel lamination, rare ripple cross-lamination and undulating lamination. These contain thin interbeds of siltstone with sandstone streaks and laminae.

Higher in the sequence, to the north of Staddon Point, the sandstone is thickly bedded with intraclast breccias and rare thin siltstone beds. The sandstone beds show parallel lamination, trough cross-bedding and current ripples, and the breccias, on erosion surfaces, comprise clasts of red and grey-green mudstone and subordinate sandstone up to 0.1 m long. A section just north of Staddon Harbour [487 508] comprises parallel-laminated, fine-grained sandstone, passing upwards into thin beds of very fine-grained sandstone with interbeds of siltstone, erosively overlain by massive and cross-bedded channel sandstone. Northwards to Jennycliff Bay, the cliff section in the Staddon Formation is poorly accessible, but where observed between Ramscliff Point [487 514] and Wyatts Way [488 516] it shows medium- to thick-bedded (up to 1.0 m), fine-grained sandstone with parallel lamination and thin mudstone partings, comparable to the sequence about Staddon Harbour. The northern boundary of the formation [4905 5180] is an undulating, gently inclined thrust fault: the sandstone succession is thrust northwards over steeply inclined beds of the Saltash Formation.

At Cawsand Bay, the Staddon Formation forms the cliff section between the Sandway Cellar area [441 551] and the Mount Edgcumbe Fault at Redding Point [4595 5185]. On structural grounds, this sequence is high in the formation although its precise position is unknown. The thickness of rock present in a family of moderately to steeply inclined, northward-verging, asymmetrical, small-scale folds is of the order of several tens of metres. The lowest part of the sequence is between Hooe Lake Point [449 513] and Picklecombe Point [506 515] and higher parts are present towards Sandway Cellar. The section shows medium- to coarse-grained sandstone beds and composite units (up to 11 m thick) and planar fine sandstone beds up to 0.5 m thick. Within these, there occurs pale grey, yellow and green siltstone as partings and sporadic thick beds, showing local red and purple staining. The massive, coarser grained sandstone beds (up to 4 m thick) show parallel, tabular and low-angle cross-bedding and lamination. They generally have erosional bases which cut down to 2 m into the underlying rocks. Where channel forms are present, layers of mudstone intraclasts are abundant, and slump folding directed towards the thickest parts of the channel fills is common. The fine-grained sandstone beds show parallel lamination, fine cross-lamination (Plate 3) with climbing sets and ripple marking. The siltstone beds contain wisps, lenses, and laminae of fine sandstone, commonly showing soft sediment deformation structures, and phacoids of slumped fine sandstone. Shrinkage cracks are sporadically present.

## BIOSTRATIGRAPHY

Fossil casts resembling *Spirifer hystericus* and *Chonetes* were recorded during the earlier survey (Ussher, 1907) but no new macrofossils have been discovered since. Burton and Tanner (1986) attributed a macrofauna from near Trevelmond [2046 6351] to the upper part of the early Emsian and to the Staddon Grits (Formation) but the presence of the formation at this locality has not been confirmed.

Miospore assemblages retrieved from the basal part of the formation in Bovisand Bay have indicated an Emsian age (Molyneux, 1990a). The presence of *Emphanisporites schultzii* is recorded, and this is restricted to the upper part of the Emsian (Richardson and McGregor, 1986). This accords with Dean's (1992) analysis of spores from Staddon Point indicating a late Emsian age. Some

**Plate 3**   Climbing ripples in a thick sandstone bed of the Staddon Formation in the cliffs [4435 5326] of Cawsand Bay, looking northwards (GS511). Hammer is 0.30 m long.

samples from stratigraphically below the formation in Bovisand Bay have yielded a mid(?) Emsian miospore assemblage (Dean, 1989a, 1992). Late Emsian spores have been identified in samples from above and north of the northern contact, within the Saltash Formation, on both east and west sides of The Sound (Dean, 1992). Thus the formation appears to be essentially of late Emsian age.

DEPOSITIONAL ENVIRONMENT

The upward-coarsening and upward-thickening features, observed in the basal part of the formation at Bovisand Bay, represent a transition to a higher energy regime from the shallow marine setting of the Bovisand Formation. The succeeding sequence to Staddon Point has been interpreted by Humphreys and Smith (1988) as showing deposition in standing water on a coastal plain or in a shallow marine embayment (compare with Muir and Rust, 1982; Plint, 1983) The 'swaley' cross-stratification reflects an environment where the embayment was open to storm waves.

The overlying sequence north of Staddon Point indicates fluvial deposition because the poorly sorted intraclast breccias associated with the channel sandstone units are typical of fluvial channel lag deposits (Clifton, 1973; Nemec and Steel, 1984) and the absence of wave-formed structures and bioturbation indicate a nonmarine setting. The section to the north of Staddon Harbour was interpreted by Pound (1983) as an upward-coarsening prodelta sequence but this was reinterpreted by Humphreys and Smith (1988). They considered that the lower part of that sequence represents planar beds of shallow channels in the upper flow regime, or proximal overbank sheet floods (compare with Tunbridge, 1981b; Friend et al., 1986), overlain by distal overbank and levee deposits. The overlying channel deposits are interpreted as secondary channels or crevasse splays. The higher parts of the sequence reflect a similar pattern of fluvial sedimentation.

At the northern overthrust boundary of the formation in Jennycliff Bay, the sandstone beds immediately above the thrust (Plate 4) have been interpreted by Pound (1983) as wave-generated bar deposits that built up above fair weather wave base.

This section at Staddon reflects an overall regression with shoreface progradation, followed by the development of a coastline with shallow lagoons or marine embayments, and culminating in a fluvial regime with deposition from sheetfloods and low relief channels.

At Cawsand Bay, the thick sandstone beds are fluvial, as indicated by their channel forms, intraclast breccias, and laterally accreted bedforms and current structures. The absence of marine influences together with the parallel lamination of some sandstone beds is suggestive of ephemeral stream/terminal fan facies models (Tunbridge, 1981a, b, 1984; Olsen, 1987). The abundance of cross-bedding features indicates that relatively deep, lower flow regimes were important, possibly reflecting high discharge. The finer grained rocks show evidence of shallow water deposition (wave ripples) and exposure (desiccation cracks). The observed soft sediment deformation structures commonly occur in shallow playas or lakes (Clemmenson, 1979; Tunbridge, 1984).

This sequence comprises overbank and shallow water deposits and fluvial sandstone. Compared with the features observed to the east of The Sound, there are thicker channel deposits, and more abundant trough cross-bedding structures and beds showing lateral accretion. This led Humphreys and Smith (1988) to propose that deposition occurred further inland at Cawsand Bay. They inferred a palaeoslope from west to east and the establishment of a substantial fluvial coastal plain during the deposition of the formation.

## TAMAR GROUP

### Saltash Formation

The Saltash Formation (Figure 16, Table 1) is a new lithostratigraphical term. It consists predominantly of grey

**Plate 4**
Gently inclined
thrust between
the Staddon
Formation
(hanging wall)
and the steeply
inclined beds
of the
Jennycliff
division of the
Saltash
Formation
(footwall) in
Jennycliff Bay
[4907 5186]
(GS512).
Hammer (left
centre) is
0.30 m long.

mudstone that extends stratigraphically from certainly the late Emsian to the Tournaisian, and possibly the Viséan. It occurs in the Looe and South Devon basins, between the crop of the Staddon Formation to the south and the fault-bounded southern margin of the Tavy Basin sequence to the north. Within this tract, it interdigitates with the Plymouth Limestone Formation and the Torpoint Formation; these formations reflect differing sources and areas of sedimentation within the basin and rise complex of south Devon and central Cornwall.

The formation includes the Jennycliff Slates (Hobson, 1976; House et al., 1977; Pound, 1983), termed the Jennycliff Slate Formation by Chandler and McCall (1985), which crops on the eastern side of The Sound between the Staddon Formation to the south and the Plymouth Limestone Formation to the north. This part of the succession extends from the late Emsian (see p.47) into the Eifelian stage of the Middle Devonian (Orchard, 1977). The Plymouth Limestone is absent at crop to the west of Mount Edgcumbe Farm [444 527], due to a combination of stratigraphical, structural and topographical factors, and here the Saltash Formation, although faulted by Millbrook Lake, extends northwards to include Upper Devonian rocks.

Over the remainder of the crop in the eastern half of the district, the Saltash Formation interdigitates on all scales with the purple and green sediments of the Torpoint Formation. Near the Tamar valley, the Saltash Formation has extensive crops in strike belts in the vicinities of Saltash and Landulph where sequences date from the Mid Devonian to the Late Devonian and possibly to the Dinantian (Early Carboniferous) in the latter case.

The Torpoint Formation dies out westwards across the district and there the Saltash Formation constitutes the whole succession, of late Emsian to Famennian age. In the Liskeard area, this succession was separated into the Tempellow Slate, Rosenun Slate and Milepost Slate formations by Burton (1974) and Burton and Tanner (1986) but these terms have been discontinued because of their lack of lithostratigraphical integrity.

The Saltash Formation consists predominantly of dark grey and grey, slaty mudstone and silty mudstone but locally there are significant subordinate lithologies, notably volcanic rocks, limestone and sandstone. There are five informal geographical and stratigraphical subdivisions of the formation in the eastern part of the district. These subdivisions cannot be used in the extreme west, where the Torpoint Formation dies out, and here a sixth undifferentiated subdivision is named. The distribution pattern is complicated not only by the north-west-trending strike-slip faults, which cut the crops obliquely, but by the variable structure, in terms of folding and faulting, along the strike of the belt. All of these factors are a function of a complex depositional basin configuration, further explained in Chapter 11.

The informal geographical subdivisions (Figure 16) are Jennycliff, Insworke, Antony, Saltash, Landulph and St Keyne. These divisions are described separately below. The St Germans Tuff Member lies within the Insworke division and the Wearde Sandstone Member is in the Saltash division.

### Jennycliff division

This subdivision is so named in order to retain the usage as applied to the Jennycliff Slates and Jennycliff Formation of previous authors (see above). The well

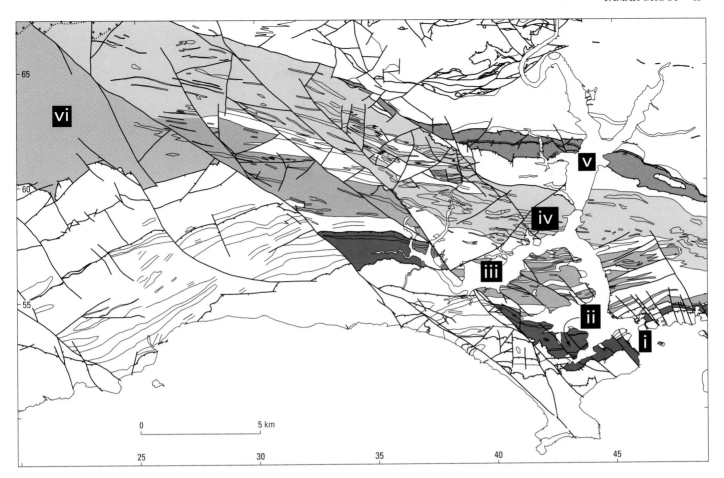

**Figure 16** Geographical subdivisions of the Saltash Formation.

i. Jennycliff division;   ii. Insworke division;   iii. Antony division;   iv. Saltash division;   v. Landulph division;
vi. St Keyne division

exposed and largely continuous Jennycliff Bay and Batten Bay section, between the Staddon Formation and the Plymouth Limestone Formation, lies mainly in the adjoining Ivybridge district (see BGS, 1996a). The section on the western side of The Sound is less complete and substantially disrupted by north-west-trending strike-slip and related faults, but all elements of the Jennycliff sequence are present.

To the west of The Sound towards Millbrook, the thrusts which bound the crop of the Plymouth Limestone Formation merge and north-west-trending faults juxtapose rocks of differing structural levels so that the formation is absent there. The westward continuation of the fault, which forms a northern boundary to the Jennycliff division, is interpreted to lie along Millbrook Lake. This fault is thought to be a thrust although, in the area immediately west of Cremyll, boundaries with the Plymouth Limestone Formation are mapped as steep faults. In this area and westwards, the Staddon Formation is thrust over the Saltash Formation. This feature is particularly marked to the north-west of Millbrook, adjacent to the Cawsand Fault, where the flat thrust base of the Staddon Formation cuts more than one kilometre northwards across strike.

This division comprises mainly dark grey to grey, slaty mudstone and fine siltstone. In the lowest 200 m of the sequence, north of the faulted contact with the Staddon Formation at Inner Redding [4595 5190] and beneath the overthrust in Jennycliff Bay [491 519], sandstone and siltstone are common. The sandstone is generally thinly bedded, but there are a few packets, exceptionally up to 6 m in thickness, comprising sandstone beds, up to 0.8 m thick, and thin silty mudstone partings. Individual thin beds show normal grading, sharp bases with gutter casts, parallel lamination in the lower part, low-angle cross-lamination in the upper part, and diffuse tops. The thicker beds show hummocky cross-stratification and sedimentary ripples (as well as tectonic rippling). Some laminae and thin beds of siltstone and very fine-grained sandstone, which weather reddish brown and purplish brown, are also common in the lowest few hundred metres of the division.

Within the mudstone and siltstone of the low part of the division, burrows, particularly of *Spirophyton*, are abundant (Humphreys and Knox, 1987). Layers of disarticulated brachiopods shells and crinoid columnals, and laminae and beds of coarse bioclastic limestone, up to 0.1 m thick, some of which show rippling, are present in the lower parts of the sequence. The abundance of sandstone decreases upwards through the division whereas the proportion of limestone increases. At Dunstone Point, in the Jennycliff Bay sequence, a thrust-bounded packet of thin limestone beds, some 5 m thick, extends westwards on to the margin of the district [4885 5267] and a similar occurrence forms The Ravens Cliffs [4585 5250], in the Mount Edgcumbe Park coastal section. These tabular and lenticular beds have erosional bases and gradational tops with intercalated grey calcareous slaty mudstone. The limestone is grey and dark grey crinoidal packstone [for example E 67235] with basal lags containing unabraded crinoid, brachiopod, coral, stromatoporoid and bryozoan fragments.

Volcanic rocks form a major part of the highest part of the sequence, south of the Plymouth Limestone crop in Batten Bay on the east of The Sound, and between Barn Pool [455 529] and Raveness Point [4596 5238] on the west of The Sound. It is probable that volcanic rocks, which form the southern half of Drake's Island [469 528], also belong to this part of the sequence. In Batten Bay, the rocks are predominantly bedded tuffs, whereas to the west units of massive, hyaloclastitic basalt are interbedded with the sequence. The thrust, which separates the Jennycliff sequence of The Sound from the structurally underlying and stratigraphically younger Plymouth Limestone Formation, is located within the tuffs in Batten Bay, and similarly on Drake's Island.

BIOSTRATIGRAPHY

The probable late Emsian to Eifelian (*Icriodus corniger* zone) age range of the Jennycliff Slates, reported by Orchard (1977), was based largely on microfaunas from thin beds of crinoidal limestone, low in the division, and on conodonts obtained from limestone at Dunstone Point (Orchard, 1978a). During the course of this resurvey, palynomorph studies have confirmed a late Emsian age for the lowest part of the division (Molyneux, 1990a; Dean, 1992). The early Eifelian age of conodonts from the limestone beds at Dunstone Point has been confirmed, and the low to mid Eifelian age (*Polygnathus costatus costatus* to ?*Tortodus kockelianus australis* zones; see Table 10) of limestone beds at The Ravens Cliffs has been determined (Dean, 1994a: personal communication, 1996). Higher in the division, limestone beds, between tuff and hyaloclastite units in Batten Bay, have yielded conodonts with late Eifelian and early Givetian affinities but they are not diagnostic (Dean, 1993).

DEPOSITIONAL ENVIRONMENT

The occurrence of *Spirophyton* near the base of the Jennycliff division is significant because it is a trace fossil documented from shallow marine environments (Miller and Johnson, 1981; Goldring and Langenstrassen, 1979) which is consistent with the encroachment of the sea over the fluvial Staddon Formation (Humphreys and Smith, 1988). The gutter casting and hummocky cross-stratification of the sandstones in the lower part of the division are compatible with the recycling of sand from the drowned coastal fluvial plain by storm activity. The thinner sandstone beds with diffuse tops are characteristic of deposits formed from suspension near wave base. The mudstone which forms the major part of the division, probably represents quieter water sedimentation, disturbed only by bioturbation, and periodic scouring currents which introduced bioclastic debris. The features of the limestone beds higher in the formation are consistent with deposition from short-lived high energy currents; the unabraded lag components indicate minimal transport and possibly represent autochthonous winnowed assemblages. The packstone textures may demonstrate incomplete winnowing or postdepositional infiltration of carbonate mud. The bedding structures may represent a turbidite origin but are also compatible with storm deposition on a marine carbonate shelf.

Insworke division

This division comprises rocks immediately north of the crop of the Plymouth Limestone Formation in central Plymouth, those extending between the Insworke peninsula and Antony east of the Cawsand Fault, and those through the St Germans area between the Cawsand and Portwrinkle faults (Figure 16). This division has a steep-faulted, northern boundary with the Torpoint Formation and younger parts of the Saltash Formation in Plymouth. Westwards, this contact is a thrust, just north of the St Germans Tuff Member between the Cawsand and Portwrinkle faults.

The division is exposed in the estuary cliffs by Millbrook Lake and St Johns Lake and comprises predominantly grey, slaty, silty mudstone with siltstone. Subordinate lithologies include laminae and thin beds of fine-grained sandstone that are commonly calcareous and macrofossiliferous, limestone beds and volcanic rocks. Bioturbation is common and clearly seen in laminated sequences. Siltstone and silty fine-grained sandstone form laminae, lenses and beds that are generally thin, but exceptionally up to 0.3 m thick. These are locally iron-stained and commonly contain dispersed comminuted brachiopod shell debris. The limestone is present as nodular horizons, as lenses and as continuous beds, up to a maximum thickness of 0.5 m. Where the thicker beds are interbedded with calcareous slaty mudstone, they form packets up to 15 m thick (as near Insworke Point [436 534]) towards the northern boundary of the division. The limestone beds show many of the characteristics of turbidites, such as normal grading, massive lower parts with coarse basal lags, parallel and cross-laminated divisions.

Lithostratigraphical and biostratigraphical evidence indicates correlation of part of this sequence with grey slaty mudstone and limestone in central Plymouth, seen only in boreholes just to the north of the crop of the Plymouth Limestone Formation. In these, the limestones are thin to thick beds and normally graded from coarse to fine grained. Lenses and thin beds of sedimentary breccia composed of fragments and blocks of stromatoporoids and corals also occur.

**Table 10** Devonian conodont, ammonoid, miospore and ostracod biozonal schemes (after Ziegler and Sandberg, 1990; Oliver and Chlupac, 1991; Streel et al., 1987 and Gooday, 1973).

| Series | Stage | Conodont biozone | Ammonoid biozone | | Miospore biozone | Ostracod biozone |
|---|---|---|---|---|---|---|
| UPPER DEVONIAN | FAMENNIAN | Siphonodella praesulcata | WOCKLUMERIA | Acutimitoceras carinatum / Cymaclymenia euryomphala / Wocklumeria sphaeroides / Kalloclymenia subarmata | Retispora lepidophyta–Verrucosisporites nitidus | Maternella hemisphaerica– |
| | | Palmatolepis gracilis expansa | | | Vallatisporites pusillites–Retispora lepidophyta | |
| | | Palmatolepis perlobata posters | CLYMENIA | Piriclymenia piriformis / Ornatoclymenia ornata / Progonioclymenia acuticostata / Protoxclymenia serpentina | Rugospora flexuosa–Grandispora cornuta | Maternella dichotoma |
| | | | PLATYCLYMENIA | Platyclymenia annulata / Prolobites delphinus / Pseudoclymenia sandbergeri | | Richterina (?Fossirichterina) intercostata |
| | | Palmatolepis rugosa trachytera | | | | serratostriata/intercostata Interregnum |
| | | Palmatolepis marginifera | CHEILOCERAS | Sporadoceras pompeckji | | Entomozoe (Richteria) serratostriata- |
| | | Palmatolepis rhomboidea | | Cheiloceras curvispina | Auroraspora torquata–Grandispora gracillis | Entomozoe (Nehdentomis) nehdensis |
| | | Palmatolepis crepida | | | | Ungerella sigmoidale / Entomoprimitia splendens / reichi/splendens Interregnum / Bertillonella (Rabienella) reichi / Bertillonella (Rabienella) schmidti / Bertillonella (Rabienella) volki / Bertillonella (Rabienella) materni / Bertillonella (Rabienella) barrandei |
| | | Palmatolepis triangularis | | | | |
| | FRASNIAN | Palmatolepis gigas | MANTICOCERAS | Crickites holzapfeli | ? | cicatricosa/barrandei Interregnum |
| | | | | Manticoceras cordatum | Archaeoperisaccus ovalis–Verrucosisporites bulliferus | Bertillonella (Rabienella) cicatricosa |
| | | Ancyrognathus triangularis | | | | cicatricosa/torleyi Interregnum |
| | | Polygnathus assymetricus | PHARCICER | Keonenites lamellosus / Petteroceras feisti / Ponticeras pernai / Pharciceras arenicum / Pharciceras lunulicosta / Pharciceras amplexum | | Ungerella torleyi |
| MIDDLE DEVONIAN | GIVETIAN | Palmatolepis disparilis | | | Contagisporites optivus–Samarisporites triangulatus | |
| | | Schmidtognathus hermanni / Polygnathus cristatus | | | ? | |
| | | | | | Geminospora lemurata–Cymbosporites magnificus | |
| | | Polygnathus varcus | | Maenioceras terebratum | | |
| | EIFELIAN | Polygnathus xylus ensensis | | Maenioceras molarium | Densosporites devonicus–Granispora naumovii | |
| | | | | | ? | |
| | | | | Cabrieroceras cripsiforme | Calyptosporites velatus–Rhabdosporites langii | |
| | | Tortodus kockelianus kockelianus | | Pinacites jugleri | | |
| | | Polygnathus costatus costatus | | | | |
| | | Polygnathus costatus partitus | | | Grandispora douglastownense–Ancyrospora eurypterota | |
| LOWER DEVONIAN | EMSIAN | Polygnathus costatus patulus | | Anarcestes | | |
| | | Polygnathus seratinus | | | | |
| | | Polygnathus inversus | | Teicherticeras discordans | | |
| | | Polygnathus gronbergi | | Anetoceras | Emphanisporites annulatus–Camarozonotriletes sextantii | |
| | | Polygnathus dehiscens | | | | |
| | PRAGIAN | Polygnathus pireneae | | | Verrucosisporites polygonalis–Dictyotriletes emsiensis | |
| | | Eognathus sulcantus kindlei | | | | |
| | | Eognathus sulcatus | | | | |
| | LOCHKOVIAN | Pedavis pesavis pesavis / Ancyrodelloides delta / Ozarkodina eurekaensis / Icriodus woschmidti woschmidti | | | Breconosporites breconensis–Emphanisporites zavallatus | |
| | | | | | Emphanisporites micronatus–Streelispora newportensis | |

## St Germans Tuff Member

Within the Insworke division, a significant volcanic unit which strikes east–west through St Germans, between the Cawsand and Portwrinkle faults, has been designated the St Germans Tuff Member (Barton et al., 1993). Hereabouts, the member is up to 175 m thick and comprises variously carbonate-cemented lava fragments and hyaloclastite fragments, massive lava, and bedded lithic tuffs which are graded and show sedimentary slump structures. An hyaloclastic basalt at Trethawle [268 622], on the western side of the Portwrinkle Fault, is tentatively correlated with this member.

It has also been correlated with occurrences near Insworke of tuff, bedded hyaloclastite and hyaloclastic basalt, as well as isolated pillows and lenses of volcanic rock. The southern unit in the Southdown area [437 528] comprises thickly bedded tuffs, and the northern unit, striking between Insworke Point [436 534] and Antony, is predominantly hyaloclastic basalt and bedded hyaloclastite. On structural evidence, the southern unit appears to be stratigraphically higher but the presence of strike faults between the two units, which may then be equivalent, cannot be discounted. The northern unit has greater lateral continuity than others within the Saltash Formation, although, in the Insworke area, the unit comprises separate bodies, and between St John and Antony, the three crops represent a single unit repeated by folding.

### BIOSTRATIGRAPHY

Conodonts from a limestone cropping out to the south of the crop of the St Germans Tuff Member, in the vicinity of Polbathic [3421 5672; 3555 5690] have indicated a Givetian, early to mid *varcus* subzones, age (Dean, 1995). Determinations from limestone beds, cropping out to the north of the crop of the tuff member, by St John's Lake [4299 5338] (information from R L Austin, Southampton University, 1987; 1990), St John and St Germans (Dean, 1995) have indicated a slightly younger (*Polygnathus varcus* Zone to early *Schmidtognathus hermanni/Polygnathus cristatus* Zone) age. Some graded limestone beds from boreholes [for example at 4718 5461] in central Plymouth yielded abundant conodonts indicative of the late Eifelian *Polygnathus ensensis* to early Givetian lower *Polygnathus varcus* zones (Dean, 1993).

## Antony division

The division is named from the area in which it is best exposed, near Torpoint and the southern bank of the Lynher River. Here and between the Cawsand and Portwrinkle faults to the north-west, it interdigitates with the Torpoint Formation. In the westernmost part of the district, the Torpoint Formation dies out and a tract of monotonous grey rocks form the division, about Trerulefoot [333 588] and Bethany [322 600], where exposure is poor and geological data relatively limited.

This division comprises mainly cleaved, dark grey mudstone and grey silty mudstone, with siltstone, fine-grained sandstone and sparse beds of sedimentary breccia. Boreholes in the Devonport area proved sections of uniformly dark grey mudstone, with only sporadically developed bedding lamination, and sequences of laminated, dark, medium and pale grey mudstone and silty mudstone. A local colour variation is due to a preponderance of one or other of these lithologies. Bioturbation is apparent, particularly in the paler rocks, where burrows are filled with darker mudstone. The coarser sediments occur in the lowest parts of the division, about the Lynher River to the north. There, the mainly interlaminated silty mudstone and siltstone contain thin to medium beds of fine-grained sandstone. Many of the sandstone beds, show gradational laminated bedding, comprising thin to thick laminae, each showing normal grading. Some laminae show low-angle cross-lamination in their upper parts, others a wispy internal lamination. Bioturbation is also common in this sequence. In the vicinity of Ince Castle [395 564], thick beds of sedimentary breccia are interbedded with units of interlaminated sandy siltstone and siltstone. Supported in a matrix of silty fine-grained sandstone, there are clasts of silty-grained mudstone, iron-rich siltstone, grey sandstone, limestone (up to 1 m × 0.2 m in size), and vesicular basic volcanic rocks, that show a crude overall grading within the beds. Towards the top of the division, on the north bank of St John's Lake [438 546], the division comprises mainly silty mudstone with sporadic laminae of iron-stained siltstone and fine-grained sandstone.

### BIOSTRATIGRAPHY

In the Antony area, the stratigraphical age of the division is largely determined from the age of the interdigitated units of the Torpoint Formation (see below) as Frasnian to the north, with a transition southwards to the Famennian. On the southern side of Torpoint, ?mid to late Famennian miospores and acritarchs have been obtained (McNestry, 1993) from grey silty mudstone on the foreshore [4377 5459].

In the westernmost part of the district, the A38 road cutting [267 628] near Cartuther Vean provided probable early to mid Frasnian brachiopods, including pelagic chonetaceans *Longispina maillieuxi* and *Retichonetes armatus*, as reported by Burton and Curry (1985) and Burton and Tanner (1986); these are from rocks thought to be equivalent to this division.

### DEPOSITIONAL ENVIRONMENT

The presence of acritarchs in the grey mudstone, and the cricoconarids with deep pelagic ostracods (see p.64) in the interdigitated Torpoint Formation affirms a marine setting for the division. The gradational bedding of the sandstone beds and normal grading of the siltstone beds are interpreted as the products of turbidity currents, the former representing bottom-hugging currents with a number of lighter suspensions. The mudstone is a quiet water deposit with some thicker laminae representing distal flows. The finer lamination possibly indicates seasonal variation in hemipelagic sedimentation (compare with Goode and Leveridge, 1991). The beds of sedimentary breccia have mass-flow characteristics and their clasts represent both intra-basinal and rise sources.

## Saltash division

This division (Figure 16) occurs as a dip section at Saltash, on the western bank of the River Tamar, extending some 2.5 km southwards from Saltmill Creek to Sand Acre Bay. It is bounded to the north and south by crops of the Torpoint Formation. Westwards, the crop of the division is dextrally displaced by faults of the Lynher Valley Fault Zone, and is split by interdigitating Torpoint Formation rocks between Trematon [395 595] and Trenance [346 619]. Although the strike is generally east–west, a combination of folding and faulting takes the northern boundary of the division north-westwards, so that the division abuts the formations of the Tavy Basin to the north and west of Quethiock [313 648]. To the west of Menheniot [290 628], correlation is uncertain. In the easternmost part of the district, this division is present in the northern part of Plymouth between Manadon and Mannamead.

In the section at Saltash, the oldest rocks seen in this division lie at the core of a major overturned antiformal fold, striking east–west in the vicinity of the Tamar Bridge. The southern limb increases in dip southwards and its sequence youngs to the south. The northern limb of the antiform is near-vertical but it has a low-angle faulted contact with subjacent grey and grey-green slaty silty mudstone to the north.

This division comprises grey to dark grey, silty mudstone with laminae and thin beds of siltstone, sporadic thin beds of fine sandstone and laminae of limestone. Locally, there are sections up to a few metres thick, where lamination appears to be absent in the mudstone. Where lamination is present, varying degrees of bioturbation are clearly evident. To the north of Saltash, the rocks weather to a green colour and locally have a purplish mottling.

Volcanic rocks constitute an important part of this division. These are lavas and hyaloclastites that commonly pass into high-level basaltic intrusions. Persistent extrusive centres include that at Sawdey's Rock [364 609] where component flows and volcaniclastics are peripherally interbedded with mudstone. However, lack of persistence and uncertainty of correlation preclude the attribution of any formal stratigraphical status, although Chandler and McCall (1985) did name one of the main occurrences near Hartley [483 572] as the Compton Volcanic Member (of their Plympton Slate Formation). Some repetition of crops by folding is evident in the vicinity of Wearde [429 580] and about a major closure at Wotton [376 614]. The lack of lateral continuity is due to the nature of the volcanicity rather than the structure. These volcanic rocks are described in more detail in Chapter 8.

## Wearde Sandstone Member

Within the Saltash division, beds of medium- to coarse-grained sandstone form a laterally discontinuous but distinctive unit formerly termed the Wearde Efford Grit (Ussher, 1907) but here called the Wearde Sandstone Member. It is prominent about Wearde Quay [425 577] and Trewint [372 594] and is also recognised at Trebursey Bridge [296 649] and Trewandra [329 616] to the west. The maximum thickness of this member is only a few tens of metres; the wide crops at Wearde Quay and Trewint result from repetition due to folding.

This member has three facies components (Jones, 1993) namely i) a channelised (channel fill) sandstone facies, ii) a non-channelised sandstone facies, and iii) a mudstone-dominated facies. The first two (Figure 17a) are exemplified at Wearde Quarry [4237 5766] and on the foreshore at Wearde Quay [4252 5767] respectively and the third is recorded in several boreholes (Jones, 1993) on the periphery of the main mapped occurrence at Trewint. The channelised facies is seen in sequences, up to 25 m thick, of thickly bedded massive, fine- to medium-grained sandstone, with internal erosion surfaces, asymptotic planar and trough cross-bedding, and rare interbeds of mudstone and siltstone. The non-channelised facies comprises fine- to medium-grained sandstone in laterally persistent, parallel beds, up to 1 m thick, with thin interbeds of interlaminated siltstone and mudstone. These units typically show normal grading and more massive sandstone passes upwards into ripple cross-laminated, silty, fine-grained sandstone. The sandstone within the mudstone-dominated facies is present as streaks, lenses, thin to thick laminae and thin beds. These have sharp bases and tops and show low-angle ripple cross-lamination.

The sandstone is a poorly sorted litharenite consisting of subangular to subrounded grains (for example SDP 102). They comprise grains of monocrystalline quartz, subordinate polycrystalline quartz, potassium feldspar, multiple-twinned and perthitic sodium feldspar and chert, together with acid igneous and sparse mica schist lithic fragments (Jones, 1993). The matrix (less than 10 per cent) comprises authigenic clay minerals and carbonate.

Close to the crops of this member, there are local occurrences of lenticular and bedded breccia, as at Trehan [405 582], composed of varying proportions of locally derived volcanic detritus, blocks and fragments of grey mudstone and siltstone in a sandy matrix.

### BIOSTRATIGRAPHY

The section south of the Tamar Bridge ranges in age southwards from the Givetian to the Famennian (Dean, 1992): the cleaved mudstones below and above the Wearde Sandstone Member are respectively of Frasnian and Famennian age. A probable Mid Devonian age for rocks low in the sequence was provided by Molyneux (1990b) from the presence of spores and acritarchs *Apiculatisporis* sp., *Diexallophasis?* cf. *simplex*, *Hymenozononotriletes* spp., *Retusotriletes* sp., *Gorgonosphaeridium* sp., *Multiplicisphaeridium* spp., and *Veryhachium* spp.. Identifications of *Diducites* spp., *Grandispora echinata*, *Rugospora flexuosa*, *Spelaeotriletes* spp., and *Micrhystridium* spp. by Turner (1993), in the upper part of the sequence, indicate Famennian age. To the north-west near Landrake, acritarchs and palynomorphs obtained in the vicinity of Skeltons Park Farm [3696 6198]

**Figure 17** Graphic logs of parts of the Wearde and Trehills sandstone members.

a. The Wearde Sandstone Member (Saltash Formation) in the section between the jetty and boathouse at Wearde Quay

b. The Trehills Sandstone Member (Tavy Formation) in the quarry section at Trehills Plantation

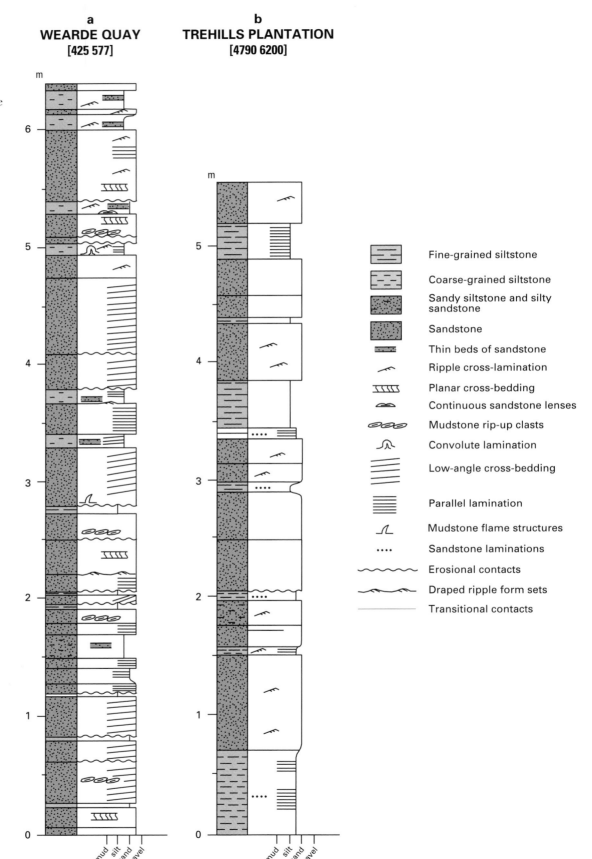

**a**
**WEARDE QUAY**
**[425 577]**

**b**
**TREHILLS PLANTATION**
**[4790 6200]**

Fine-grained siltstone

Coarse-grained siltstone

Sandy siltstone and silty sandstone

Sandstone

Thin beds of sandstone

Ripple cross-lamination

Planar cross-bedding

Continuous sandstone lenses

Mudstone rip-up clasts

Convolute lamination

Low-angle cross-bedding

Parallel lamination

Mudstone flame structures

Sandstone laminations

Erosional contacts

Draped ripple form sets

Transitional contacts

mud silt sand gravel

indicated a Frasnian to Famennian age to Dean (1989b), which was subsequently amended to Famennian (Dean, 1992) and confirmed as such by McNestry (1994a).

## DEPOSITIONAL ENVIRONMENT

The mudstone and siltstone units generally represent deposition of sediment from suspension under quiet-water conditions. Structureless siltstone without tractional structures may represent deposition from concentrated sediment suspensions. Although the acritarchs indicate a marine environment, the paucity of macrofauna may be due to a reduced salinity rather than rapid deposition.

Channelised thicker sandstones have been recognised (see above). The thick bedded and massive sandstones indicate deposition from turbidite flows within channels. The sheet sandstone beds of the non-channelised sandstone facies are interpreted as the product of unconfined flows also emplaced by turbidity currents. The massive sandstone beds represent high-density turbidites with rapid rates of deposition and those beds with sedimentary structures throughout result from low-density tractional currents with low rates of sedimentation. The non-channelised facies represents local depositional lobes (Jones, 1993). The thin laminae and beds of coarser-grained clastics in the mudstone are the products of more lateral or distal, dilute turbidity currents (compare with Collinson et al., 1991).

The Wearde Sandstone Member is present sporadically along strike over nearly 20 km in this district and also occurs further to the east. It locally forms thick channels and associated lobes with very limited lateral extent and these are associated with occurrences of extrusive volcanic rocks. This may indicate that sandstone deposition took place on an uneven topography, where floods of sand were channelled between contemporaneous volcanic piles.

## Landulph division

This northerly division of the formation is bounded by faults, principally thrusts. It is best exposed on the banks of the River Tamar near Warleigh Point and its crop extends westwards through Landulph, Hatt, and the Lynher Valley Fault Zone, to Goodmerry Farm [336 647] between Pillaton and Quethiock. There it attenuates along the Ludcott Fault at the margin of the Tavy Basin deposits. Eastwards, the crop passes through Crownhill Fort [488 592], on the eastern margin of the district.

The division is stratigraphically and structurally complex with internal faulting, including thrusting. On the west bank of the River Tamar, bedding/cleavage relationships and facing indicate that the sequence youngs to the south, except in two medium-scale folds. However, the pattern of younging is not reflected by the biostratigraphy of the rocks (see below) which occur in at least four fault-bounded subdivisions.

The most complete section is on the west bank of the River Tamar but the structurally lowest unit, to the north, is only seen on the east bank. A bedded chert lies against the hanging wall of the gently southerly inclined Crownhill Thrust that bounds the division. A black and dark grey lamination is apparent in parts of the chert in which radiolarian tests are abundant (E 3152). This chert does not extend to the west side of the River Tamar, because of faulting.

The overlying unit is predominantly medium grey, cleaved, silty mudstone, which locally comprises laminae of darker grey mudstone and paler grey silty mudstone in couplets. In the extremities of the section on the west side of the river, red- and brown-stained siltstone laminae and lenses are common locally. In the central part of the section, just south of Parsons Quay [4371 6159], limestone is sparsely present as laminae, lenses and more continuous beds up to 0.1 m thick.

The succeeding unit to the south comprises mainly dark grey and grey chert and siliceous slaty mudstone with some black slaty mudstone, several tens of metres thick. It forms a promontory some 150 m north of Neal Point, but is only exposed east of the River Tamar, on the northern shore of Tamerton Lake [4470 6090] in minor fault-bounded exposures. The northern boundary of the siliceous rocks on the west side of the River Tamar is a steep fault, with a breccia 1 m thick. A few metres to the north of here, purple, green and grey mottled, cleaved, silty mudstone occurs in the hanging wall of a southerly inclined low-angle fault: the structural situation of this lithology is uncertain. The southern boundary of the cherty unit is a moderately inclined fault (Figure 1). The breccia and fault wall of this structure provide evidence of early thrust and later dip-slip movements.

The southernmost and highest unit comprises grey silty mudstone, finely laminated in part with, to the north, thin beds of graded sandstone that are locally regularly spaced. Some finely bioclastic and pelloidal limestone (E 67231), in lenses and thin-graded beds up to 0.1 m thick, is present throughout. Limestone is more abundant, in sequences several metres thick, at Neal Point [4358 6121] and at a fault-bounded exposure in the adjacent cove [4354 6122] to the south-west. At both localities, beds of limestone up to 0.1 m thick are interbedded with thin, cleaved, grey, calcareous mudstones. At Neal Point, the limestone weathers pale pinkish brown and the thicker beds comprise irregular, nodular laminae (Plate 5). In the cove, the grey limestone is normally graded from medium- to fine-grain size (E 67228) and beds have laminated, or locally cross-laminated tops. Within this sequence there is a 'raft' of limestone, 3.0 m × 0.7 m, which forms a dislocated bed with a coarse crinoidal base (E 67229), a massive medium-grained central part and a fine-grained laminated top (E 67230). The raft is blunt-ended and shows an extension of the upper, laminated part which is disharmonically folded back on itself. The early cleavage crosses this folded part obliquely, indicating plastic deformation at the time of incorporation.

To the west of the Tamar valley, the sequence is bounded by a north-west-trending fault but other facets of the division are seen to the west and east. West of Neal Point, basic tuffs have small crops towards Moditonham and also a purple and green slaty mudstone forms a discontinuous unit. The subdivision has a southern boundary which is locally gently inclined, following the contours, on the steep, northward facing slope [405 610], to the south of Botus Fleming.

On the eastern side of the River Tamar, limestone is present in the valley which extends east-south-eastwards from Budshead Creek [462 601]. Here, the southern boundary of the division is a flat-lying thrust, oblique to the strike, which also bounds an outlier of Torpoint Formation above the Landulph division, near Whitleigh [475 598].

BIOSTRATIGRAPHY

The strata which fall in this division were attributed in large part to the Upper Devonian by Ussher (1907) mainly because of the occurrence of *Styliola* and *Posidonomya venusta* in dark [grey] slates near Botus Fleming. He attributed the chert which occurs between Warleigh Point and Whitleigh to the Lower Culm Measures (Dinantian) on the basis of affinity to other cherts of established age elsewhere in the region. He had reservations about the 'coralline limestone' at Neal Point which he thought might lie near the top of the Middle Devonian. Later, Matthews (1962) discovered conodonts in this section which he referred to the late Eifelian whilst recognising the presence of *Polygnathus varca (varcus)*, a zone fossil for the Givetian. The latter age appeared to be affirmed by Orchard (1978) who obtained conodonts from the section at the nearby cove and from a locality 40 m north of Neal Point, which he thought could be attributed to the Givetian (*Icriodus obliquimarginatus* Zone).

During this resurvey, no further information on the age of the chert unit at Warleigh has been obtained. The northernmost unit on the west bank of the River Tamar has only yielded a microfauna and although Dean (1992) tentatively suggested a late Famennian age, diagnostic forms have been retrieved from only two samples, one 40 m north-north-east of Parsons Quay [4372 6163] and the other, 140 m south-south-west of Parsons Quay [4363 6149]. The former yielded a diverse miospore population that included *Raistrickia variabilis*, *Dictyotriletes submarginatus*, *Diducites versabilis*, *Vallatisporites pusillites*, *V. verrucosus*, *Grandispora cornuta*, *Rugospora flexosa*, *Auroraspora hyalina*, *A. macra*, *A. asperella* and *Retispora lepidophyta*. Owens et al. (1993) concluded that these and other elements (Plate 6) indicate a late Famennian (Fa 2d) to early Tournaisian (Tn 1a) age (uppermost Devonian). The latter sample contained a variety of miospores (McNestry, 1993) including *Emphanisporites schulltzii* indicative of an Emsian age.

Palynomorphs obtained from the equivalent dark grey slates on the east bank of the River Tamar between the Crownhill Thrust and Tamerton Lake were identified as late Famennian to Early Carboniferous forms by Dean (1991, 1992): one sample at Warleigh Point was more specifically assigned to the lower Tournaisian (Tn 1 or Tn 2).

The central cherty unit of the western section on the River Tamar has not yielded a definitive fauna but palynomorphs from dark grey, siliceous slate on the north bank of Tamerton Lake [4472 6086] have indicated a probable Tournaisian age to Dean (1992).

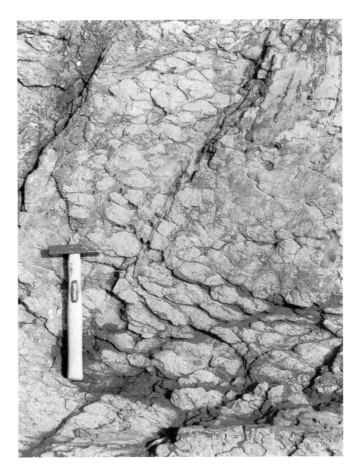

**Plate 5** Nodular limestone (Famennian) at Neal Point [4357 6123]. Viewed normal to bedding dipping moderately southwards towards the viewer (GS513). Hammer is 0.30 m long.

The southernmost unit provided conodonts from limestone and palynomorphs from silty mudstone, with similar results. Two limestones in the northern part of the unit contained conodonts of Givetian age, the Middle *Polygnathus varcus* Subzone at 110 m north of Neal Point [4360 6136] (Dean, 1994b) and *P. varcus* Zone to *Schmidtognathus hermanni/Polygnathus cristatus* Zone at 80 m north of Neal Point (Dean, 1994b). The survival of delicate elements and only slight abrasion of the conodonts indicated minimal reworking (Dean, 1994b). Limestone, in the stream bed between Tamerton Lake [460 604] and the eastern margin of the district, has yielded conodonts of uppermost Givetian age (Dean 1993; Bultynck, personal communication, 1996).

At a higher stratigraphical level, samples from Neal Point and from 50 m north of there have yielded Famennian, principally Upper *Palmatolepis crepida* Zone, assemblages (Dean, 1994b; additional information from R L Austin, Southampton University 1990, and P Bultynck, Royal Belgian Institute of Natural Sciences, 1996; see Plate 7). However, a conodont fauna from the limestone raft (see above) in the nearby cove indicated

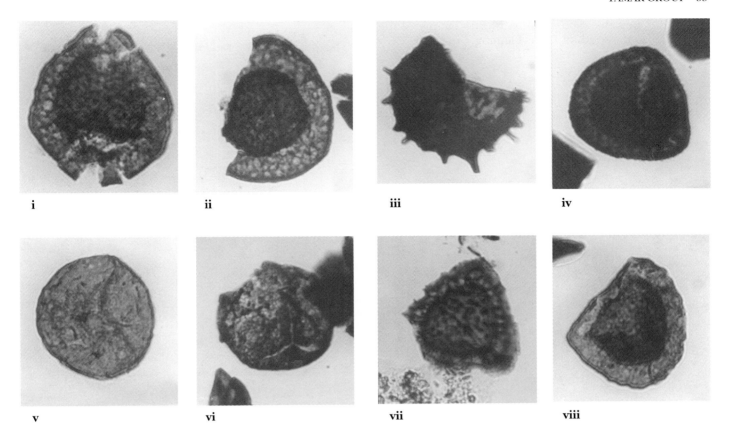

**Plate 6**  Palynomorphs from the Saltash Formation; Landulph division (GS514).

i     *Retispora lepidophyta* (Kedo) Playford, 1976, Tournaisian, Tn1a–Tn1b
ii    *Retispora lepidophyta* (Kedo) Playford, 1976, Tournaisian, Tn1a–Tn1b
iii   *Raistrickia variabilis* Dolby and Neves, 1970, Tournaisian, Tn1a–Tn3c
iv    *Spelaeotriletes* sp., ?Frasnian–Westphalian
v     *Plicatisporites* sp., Tournaisian, Tn1a–Tn3c
vi    *Retusotriletes simplex* Naumova, 1953, Mid-Late Devonian, Emsian–Famennian
vii   *Vallatisporites* cf. *pusillites*, Tournaisian, Tn1a–Tn1b
viii  *Auroraspora hyalina* (Naumova) Streel, 1974, Frasnian–Famennian, Tn1b

All specimens enlarged × 500 and selected from BGS sample number MPA 39310 from
40 m north-north-east of Parson's Quay [4372 6163]

an early Frasnian age to Austin (personal communication, 1990) and a Givetian age to Dean (1994b).

An analysis of palynomorph assemblages retrieved from slaty mudstone of this unit led Dean (1991, 1992) to propose a Givetian/Frasnian age for a sample from 110 m north of Neal Point, and Givetian/Famennian and possible Frasnian ages for two samples from the cove. The Frasnian age was based on acritarchs because the last two samples contained spores of varying age ranges that reflected substantial reworking.

The nodular limestone at Neal Point has yielded only poorly preserved cephalopods, none of which have proved identifiable (Riley, 1996). Cricoconarids (*Styliolina*) and tentaculitids are present at various levels through this division (Wilkinson, 1993c).

In summary, the radiolarian chert below the thrust at the northernmost end of the section has an affinity to Viséan examples (see Ussher 1907). The sequence above,

up to the higher cherty unit, is essentially uppermost Devonian (late Famennian–early Tournaisian). There is no structural evidence to explain the presence of the Emsian palynomorphs, even in a condensed sequence. Reworking, as found elsewhere, may provide one explanation for this occurrence. The cherty unit may form a continuation of a younging southwards sequence if the Tournaisian age (Dean, 1992) is correct. The uppermost sequence, thrust over this unit, ranges in age from the Givetian to the late Famennian, although a Frasnian fauna has not been identified in this apparently continuous sequence.

DEPOSITIONAL ENVIRONMENT

A marine environment is confirmed by the presence of acritarchs, cricoconarids and tentaculitids in the cleaved mudstones and of cephalopods in the nodular limestone of Neal Point. The majority of the limestone beds, with

**Plate 7** A selection of Devonian conodont Pa elements, in various states of preservation (GS515). Species ranges based on Austin et al. (1985) and Ziegler and Sandberg (1990). The specimens illustrated are housed in the Palaeontological Collections of the BGS, Keyworth, and registered in the series MPK10101–MPK10113.

i  *Caudicriodus curvicauda–C. celtibericus* (Carls and Gandl, 1969) group, × 56, MPA40584(10), Long Sands cliffs [3789 5314]. Species group range: *kindlei* to *gronbergi* zones. MPK10101

ii  *Polygnathus costatus costatus* Klapper, 1971, × 40, MPA37674(1), Faraday Road, Plymouth [4980 5420]. Species range: *costatus* to *australis* zones. MPK10102

iii  *Icriodus amabalis* Bultynck and Hollard, 1980, × 44, MPA40588(12), Ravens Cliff [4585 5252]. Species range: *costatus* to *enensis* zones. MPK10103

iv  *Icriodus struvei* Weddige, 1977, × 52, MPA37674(4), Faraday Road, Plymouth [4980 5420. Species range: c. *costatus* to *ensensis* zones. MPK10104

v  *Polygnathus trigonicus* Bischoff and Ziegler, 1957. × 32, MPA37680(1), Macadam Road Quarry, Plymouth [4936 5392]. Species range: *australis* to *ensensis* zones. MPK10105

vi  *Polygnathus xylus ensensis* Ziegler, Klapper and Johnson, 1976, × 39, MPA376376(1), Power Station site, Plymouth [4981 5412]. Species range: *ensensis* to Middle *varcus* zones. MPK10106

vii  *Tortodus* aff. *variabilis* (Biscoff and Ziegler, 1957), × 23, MPA37683(1), ESSO Quarry, Plymouth [4925 5389]. Species range: *ensensis* to Middle *varcus* zones. MPK10107

viii  *Ancyrodella pristina* Khalymbadzha and Tschernysheva, 1970, × 32, MPA33348(17), Crownhill [4930 5861]. Species range: Early *falsiovalis* Zone. MPK10108

ix  *Palmatolepis minuta minuta* Branson and Mehl, 1934, × 52, MPA40906(25), Neal Point [4357 6123]. Species range: Late *triangularis* to Late *trachytera* zones. MPK10109

x  *Palmatolepis quadrantinodosalobata* Sannemann, 1955, × 37, MPA40905(14), Neal Point [4357 6121]. The posteriorly directed lobe is due to deformation. Species range: Early *crepida* to Early *rhomboidea* zones MPK10110

xi  *Polygnathus nodocostatus* Branson and Mehl, 1934, group, × 52, MPA40906(16), Neal Point [4357 6123]. Species range: Early *crepida* to Latest *marginifera* zones. MPK10111

xii  *Palmatolepis glabra pectinata* Zigler, 1962, × 37, MPA40906(26), Neal Point [4357 6123]. Species range: Latest crepida to Latest marginifera zones. MPK10112

xiii  *Palmatolepis marginifera marginifera* Helms, 1959, × 50, MPA40906(31), Neal Point [4357 6123]. Species range: Early to Latest *marginifera* zones. MPK10113

Bouma (1962) sedimentary features, are turbidites, and the raft of limestone in the cove near Neal Point is evidently slumped. The admixing of palynomorphs, noted by Dean (1992), and variable age determinations by differing authors (McNestry, 1994b), are probably explained by reworking in the source areas of the turbidites and intermixing during emplacement. The chert and cephalopodic limestone indicate relatively quiet pelagic conditions, the latter lithology in a rise slope setting. The source of the limestone, here termed the Landulph High, was probably to the north of this area because limestone is generally absent from rocks of similar age to the south.

### St Keyne division

This name is given to the undifferentiated Saltash Formation which lies to the west of the Portwrinkle Fault. Poor exposure and the presence of a strong secondary cleavage transposition fabric generally preclude observation of bedding relationships or details of sedimentary structures. Hence no direct correlation can be made with the other divisions described to the east. There are, however, broad changes of bulk lithology across strike from south to north, reflecting those to the east, but they do not merit a formal subdivision. The biostratigraphical subdivision proposed by Burton and Tanner (1986) for this area cannot be sustained on lithostratigraphical grounds.

The southern part of the crop consists mainly of grey siltstone with subordinate cleaved silty mudstone and fine-grained sandstone, as wispy and continuous laminae and thin beds; limestone is sparsely developed. In the finer grained rocks, some beds, 0.1 m to 0.3 m thick, show normal grading from muddy siltstone to silty mudstone. The sandstone is locally richly macrofossiliferous with crinoidal and shell debris, and in places composite sandstone beds occur, a few metres thick. Two thick volcanic units are present to the east near the Portwrinkle Fault. The southernmost, seen at Lean Quarry [265 614], shows pillow lavas intimately associated with intrusive dolerite. That near South Treviddo [2700 6200] comprises hyaloclastitic basalt, which abuts the fault and extends some 2 km westwards. Basic tuff and volcanic breccia present near Scawns [222 628], farther to the west, are possible correlatives of this basalt.

To the north of these volcanic rocks and south of Liskeard and Dobwalls, the division comprises cleaved silty mudstone and siltstone, laminated in part. The mudstone contains sporadic laminae and thin beds of siltstone, fine-grained sandstone and limestone. The limestone is fine grained and laminated.

In the north-westernmost corner of the district, the division comprises largely slaty mudstone. Thus, within the division there is an overall fining northwards similar to that apparent between the Insworke and Saltash divisions (ii and iv) but no temporal correlation within the St Keyne division has been established.

#### BIOSTRATIGRAPHY

In the south of the district, a macrofauna from the vicinity of Rosenun [2480 6173] including *Bradocryphaeus?*

*cantarmoricus* and a trilobite *Greenops (Neometacanthus) perforatus* indicated a mid Eifelian age to Burton and Tanner (1986). Rocks between this locality and the Staddon Formation to the south were deduced by these authors to be older and correlatives of the Jennycliff (Slates) division. An extensive brachiopod fauna within a volcanic unit at South Treviddo [2700 6200] was identified as late Eifelian to Givetian by the same authors. This volcanic unit may therefore be coeval with the St Germans Tuff Member. In the Coombe area [2361 6262], north of St Keyne, thin limestone beds have yielded conodonts of probable early Eifelian to early Givetian age (Dean, 1995).

North of the volcanic rocks equated with the St Germans Tuff, macrofossils retrieved from grey cleaved mudstone near Trevelmond [2046 6351] were thought by Burton and Tanner (1986) to confirm a late early Emsian age and equivalence with the Staddon (Grits) Formation. A key element to this determination was the presence of *Pleurodictyum problematicum*, which was also identified by Burton (1972) at Laira [5000 5431] in Plymouth, in green slaty rocks now mapped within the Middle Devonian. Similarly at Lantoom Quarry [225 649], to the west of Liskeard, silty mudstone with transposition cleavage fabric was shown as Staddon facies on their map of the area. No confirmatory biostratigraphical data was obtained from these localities during the survey.

#### SUMMARY OF DEPOSITIONAL ENVIRONMENT

The grey fine-grained rocks forming the Saltash Formation form a background sedimentation for the basinal areas from the upper part of the Lower Devonian into the Lower Carboniferous. In the basin, they occur as finely laminated sequences and more homogeneous dark grey mudstone. Fine-grained sediments were also laid down on rise slopes, as possibly condensed sequences, and in a shallow shelf setting starved of coarser clastics. Bioturbation is locally common but benthic fossils are rare, except in the shallower shelf regime where their presence is largely due to storm or turbidity current transport.

## Plymouth Limestone Formation

This formation has a restricted occurrence in the southeast of the district, forming the waterfront of the city and Drake's Island in The Sound. It also crops near Cremyll and Mount Batten, on the north-east and north-west sides of The Sound respectively (Figure 18). The formation extends eastwards towards Cattewater and Plymstock in the adjoining district.

The recent resurvey, more detailed than has hitherto been available, confirms that the formation is structurally isolated (Figure 19). Previous authors recognised the structural complexity of the formation and attempted to reconcile its biostratigraphy with an interpreted structure, in order to produce a coherent local stratigraphy. Ussher assigned the 'Plymouth limestone' to the Middle Devonian, on the basis of its rich macrofauna of corals, brachiopods and gastropods (1907, pp.51–53), figured and identified earlier, mainly by

Worth (1878). A detailed investigation of the rugose corals within the limestone led Taylor (1951) to propose that two parallel crops of Middle to Upper Devonian sequences formed the limbs of an overturned fold, The northern limb through the city waterfront was thought to be gently inclined southwards and the southern limb, much steeper and inverted. Both northern and southern boundaries of the composite crop were thought to be faults. A comparison of the limestone with that of modern reefs had been made at an early stage by Worth (1888) and this analogy was used by Ussher (1907). Braithwaite (1967) suggested that the formation was essentially bioclastic limestone derived from transient patch reefs and lacked a reef framework. Thrusting along the southern boundary of the formation was invoked when he determined that the limestone had a sheet-like form and was not a faulted syncline. The synform concept was nevertheless revived by Orchard (1978a) to explain a distribution of Middle to Upper Devonian biostratigraphical ages based on his analysis of conodonts from the limestone sequences. He also differentiated areas of biogenic growth, back reef and fore reef facies within the limestone. Subsequently, evidence was presented by Chapman et al. (1984) that at least part of the northern boundary of the formation is a thrust, and by Chandler and McCall (1985) that the southern boundary to the south of Mount Batten is also a thrust.

The steep, faulted nature of the northern boundary is now established in the Sutton Harbour and The Hoe area in numerous site investigation boreholes, and field evidence indicates a similar situation at Stonehouse [4598 5440]. These are late structures however and the basic structure of the formation is probably that of a thrust

duplex modified by secondary thrusting (Figure 19, see Chapter 11). Extensional strike faults and numerous cross-cutting strike-slip faults are also recognised in the city.

## PLYMOUTH

In the Cattedown area to the east of Sutton Harbour [484 540], where there is continuous exposure in a series of road cuttings and quarries, the formation has been subdivided into the Faraday Road Member, the Prince Rock Member and the Cattedown Member, in ascending order. Only the lower member, which is readily distinguishable in boreholes and temporary excavations, has been recognised extensively in this district. There is a lateral facies change in the formation (see p.60) and the distinguishing features of the upper two members do not persist westwards.

The limestone is classified here according to the textural scheme of Dunham (1962), as modified by Embry and Klovan (1971). The limestone was subject to shearing and strain recrystallisation and, at some localities (for example Wilderness Point [456 531] and Rusty Anchor [472 536]), these features are pervasive.

### Faraday Road Member

This basal member is widely mapped through the city and its type section is along Faraday Road, Cattedown, in the adjacent Ivybridge district. Up to 150 m of strata make up the member between the faulted base of the formation and the overlying rocks. It comprises dominantly thin, dark grey, commonly foetid, crinoidal packstone intercalated with calcareous mudstone or argillaceous, fine-grained wackestone. The limestone beds have sharp, erosional bases. The packstone beds vary from tabular to lenticular, and locally exhibit normal grading and cross-stratification. Flattened burrowfills occur in the mudstones and at the bases of the limestone beds.

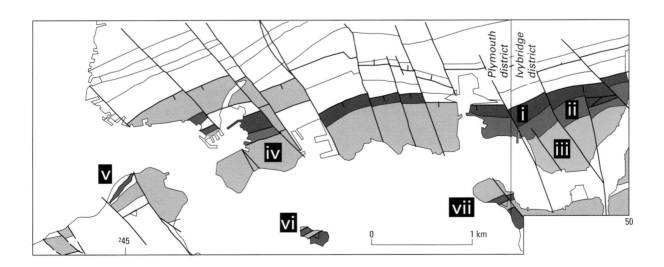

**Figure 18**   Plymouth Limestone Formation.
i Faraday Road Member;   ii Prince Rock Member;   iii Cattedown Member;   iv Plymouth Limestone Formation, undivided;   v Cremyll and Empacombe;   vi Drake's Island;   vii Mount Batten

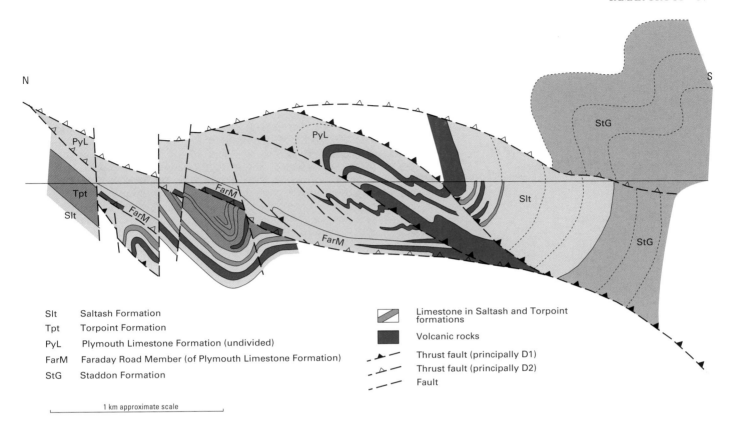

N

S

| | |
|---|---|
| Slt | Saltash Formation |
| Tpt | Torpoint Formation |
| PyL | Plymouth Limestone Formation (undivided) |
| FarM | Faraday Road Member (of Plymouth Limestone Formation) |
| StG | Staddon Formation |

Limestone in Saltash and Torpoint formations

Volcanic rocks

Thrust fault (principally D1)

Thrust fault (principally D2)

Fault

1 km approximate scale

**Figure 19**  Diagrammatic section illustrating the structure of the Plymouth Limestone Formation (not to scale). Differing relationships are apparent along strike due to variation in the location of the original ($D_1$) thrust sheets, degree of secondary thrusting, normal fault displacements on strike-slip faults, differing throws on steep strike faults and down-dip movements during orogenic extension.

Stylonodular pressure solution effects (Wanless, 1979) are widespread. The limestone locally displays richly fossiliferous, basal lags composed of rolled, but not significantly abraded, crinoid, brachiopod, stromatoporoid and bryozoan fragments. The mudstone contains scattered and commonly crushed brachiopods, corals and crinoid debris. In larger sections of the member, the thickness and frequency of the packstone beds commonly increases upwards with a concomitant reduction in mudstone content. The merging of limestone beds is also more common in the higher parts. In its type area, 50 m of grey-green, basaltic, tuffaceous mudstone forms the upper part of the member.

In this district, the Faraday Road Member is partly exposed in a quarry on the west side of Stonehouse Pool, at the southern end of Richmond Walk [459 540], where the northern and lower boundary is a steep, faulted contact with higher levels of the formation. Bedded hyaloclastite, with sporadic thin beds of coarse crinoidal packstone, passes up into medium beds of limestone with thin partings of dark grey mudstone. The succeeding grey slaty mudstone is locally calcareous and tuffaceous, with sparse thin beds of limestone.

In the wackestone beds of the section at Faraday Road, Orchard (1978) reported the disarticulated remains of the trilobites *Phacops* cf. *hefteri, Otarion* cf. *druida* and *Proetus* sp., as well as small solitary corals, atrypid brachiopods, *Pleurodictyum*, ostracods, bivalves and fragments of fenestellid bryozoans. Orchard (1978) also recorded conodont assemblages with a

late Eifelian age. However, conodonts from the lower limestone beds of the Richmond Walk section include *Icriodus corniger* indicating an age no younger than mid Eifelian (Orchard, 1978). This may indicate that part of the type section is older than presently determined.

**Prince Rock Member**

The type section is in the Prince Rock area [498 541] close to the east of this district where generally this member overlies gradationally the tuffaceous mudstone of the Faraday Road Member. It has an estimated maximum thickness of 100 m. The Prince Rock Member comprises a sequence of thin- to medium-bedded, fine- to medium-grained skeletal wackestone and packstone with stylolitic mudstone partings and interbeds. Coarser crinoidal lithologies occur widely scattered throughout the lower parts of the member and form the basal lags of graded and cross-stratified packstone beds which are common in the upper part. The limestone is typically dark grey and foetid. Thick to very thick, tabular and lenticular beds of coral, stromatoporoid and *Chetetes* boundstone and floatstone, the latter typically with a crinoidal matrix, occur dispersed throughout the member. In the upper part of the member there are lenticular beds of floatstone and rudstone, in which broken and overturned coral and stromatoporoid colonies overlie erosional and locally channelised basal surfaces. The lowest and thickest (4.4 m) boundstone occurs some 8 m above the faulted base of the member in Faraday Road. Such beds commonly exhibit gradational bases and sharp, undulating tops. Succeeding limestone beds pinch out against such tops, and this feature appears to indicate a primary depositional topography. In these beds, most of the colonial fossils appear in to be preserved in life position.

Taylor (1951) recovered corals, probably from the lowest floatstone/boundstone unit in this section, which include *Acanthophyllum* cf. *fibratum, Leptoinophyllum* sp., *Domophyllum* cf. *abreviatum* and *Amplexus* sp.. From a higher level in the member Taylor listed the taxa *Macgeea bunthi, Phillipsastraea* sp., *Acanthophyllum* sp., *Sparganophyllum* cf. *defficile, Favosites* sp., *Heliolites porosus* and *Alveolites* sp.. Conodont assemblages recorded by Orchard (1978) from the type section include *Polygnathus pseudofoliatus, P. linguiformis linguiformis*, rare *P. angustipennatus, Ozarkodina bidentata, Beodella, Neopanderodus* and *Icriodus* and are consistent with a late Eifelian age. The entry of *Polygnathus xylus* and *Icriodus regularicrescens* in beds exposed at Teats Hill Quarry [486 541], and the presence of *Tortodus* sp. *P. latus, Ozarkodina brevis, I. obliquimarginatus* and elements of the *I. expansus* group in the uppermost beds of the member in Cattedown [493 539], confirm that the age of this member ranges up into the early Givetian (Orchard, 1978).

In the type area, the Prince Rock Member gives way upwards and southwards to the Cattedown Member (BGS, 1996a) which shows a late Givetian reef and back-reef facies (Davies, *in preparation*). This member has not been recognised in this district.

**Plymouth Limestone Formation (undivided)**

In the city, west of Sutton Harbour [484 541], the Plymouth Limestone cannot be divided except where the Faraday Road Member has been identified in boreholes and in an exposure at the southern end of Richmond Walk [459 540]. Hereabouts, there is present another distinctive facies, in which stromatoporoid boundstone and crinoidal packstone/grainstone are closely associated, and which is coeval with and possibly younger than the Cattedown Member. This facies, identical to that described by Mayall (1979), occurs in two distinct tracts. The southerly tract extends from Sutton Harbour through The Hoe, and Millbay Docks [469 538] to Devil's Point. The northern belt extends westwards from the north-west-trending Millbay Fault that strikes along the eastern side of Millbay Docks, through Stonehouse and Mount Wise (see Figure 8).

Limestone occurrences to the south and west of the city waterfront, at Cremyll and Empacombe, Drakes Island, and Mount Batten, also constitute parts of the undivided formation.

The detailed conodont biostratigraphy of Orchard (1978) has demonstrated that the exposed undivided sequences range in age from mid Givetian (*Polygnathus varcus* zone) to mid Frasnian (upper *asymmetricus* subzone) in age and confirms their equivalence to the Cattedown Member farther east. Early Famennian conodonts at Western King Point, recovered from red calcite mudstones, have been interpreted as fissure fillings (Orchard, 1975).

The lithologies are well exposed in the cliffs of The Hoe, and between Devil's Point and Eastern King Point, and in numerous quarry faces including a complete section [459 542] of the northern belt by Richmond Walk. Typical sections show irregular masses of fine-grained, pale grey stromatoporoid boundstone which pass laterally and vertically into coarse, crinoidal packstone and grainstone with scattered stromatoporoids and corals. The stromatoporoid boundstone masses vary greatly in size and are up to 4 m in thickness and tens of metres across. The framework consists of recrystallised, interfingering stromatoporoid colonies of dominantly tabular

habit. Spar-filled shelter cavities are common, but crinoidal debris typically occupies the interstices between the colonies. The crinoidal parts of the section commonly exhibit crude cross-bedding and grading, and are locally pink in colour.

Beds of floatstone and rudstone with reworked and broken coral and stromatoporoid colonies and erosional bases, comparable to those in the Prince Rock Member, are also present in the boundstone and crinoidal sequences. Elsewhere, beds of fine-grained, skeletal packstone with scattered large, individual stromatoporoid colonies interdigitate with the dominant lithologies.

Dark, foetid, thin-bedded, skeletal wackestone, rich in the distinctive branching stromatoporoid *Amphipora*, also occurs in the higher parts of the Plymouth Limestone Formation exposed at Western King [461 533]. Here, and at other localities, conspicuous fissures filled with red, argillaceous calcite mudstone or breccia both cross-cut and extend parallel to the bedding in the limestones.

In the northern tract, the westernmost exposures of the Plymouth Limestone Formation are of dolomitised, crinoidal packstone seen at Mutton Cove [453 540], which yields conodont assemblages, including *Polygnathus linguiformis klapperi, P. ansatus* and *Ozarkodina brevis* of the *varcus* Zone (Orchard, 1978). The sequence exposed in disused quarry walls by Richmond Walk [4608 5441 to 4586 5409] is moderately inclined southwards, apart from an open small-scale syncline/anticline fold couplet adjacent to the bounding fault. Here, tuffaceous mudstone is overlain by dark, thin-bedded, argillaceous wackestone and packstone. Corals reported from the limestone beds include *Macgeea varians, Mesophyllum thomasi, Cylindrophyllum stonehousense, Acanthophyllum* sp., *Favosites* sp., *Alveolites* sp. and *Callopora* sp. (Taylor, 1951; Ussher, 1907). Scattered stromatoporoids are also present. Conodont assemblages include abundant *Belodella* sp., with *Polygnathus timorensis, P. linguiformis linguiformis* and *Ozarkodina brevis* and are indicative of the mid Givetian, lower *Polygnathus varcus* subzone (Orchard, 1978). Above, medium to thick beds of pale grey limestone, composed of crinoid, stromatoporoid and colonial coral debris, contain irregular blocks of purplish red, silty mudstone showing evidence of soft sediment deformation. Thicker succeeding beds comprise coarse crinoidal packstone and boundstone units dominated by in-situ tabular stromatoporoids but including massive colonies (as constituting one 4 m-thick unit [at 4586 5412]). Rudstone beds of reworked and broken stromatoporoids and corals also occur together with coquinas of disarticulated brachiopod valves. The conodont assemblage present in these beds, including *Polygnathus ansatus, Polygnathus linguiformis weddigei* and *Icriodus latericrescens latericrescens*, is of the upper *Polygnathus varcus* subzone in the late Givetian (Orchard, 1978).

In the southern tract, the cliffs from just north of Devil's Point [4594 5341] to Western King Point [4619 5324] afford a section through the youngest exposed parts of the formation. Stromatoporoid boundstone and floatstone of probable Givetian age, exposed at the north-western end of the section, give way southwards to a sequence of dark, foetid wackestone and packstone with common *Amphipora*, exposed at Devil's Point. Thin rudstone with reworked stromatoporoids, corals and brachiopods are also present in the sequence. To the south-east, between the two points, tabular stromatoporoid boundstone and coarse, crinoidal packstone/grainstone alternate with finer grained, skeletal packstone. Frasnian conodont assemblages indicative of both the lower and upper *asymmetricus* biozones have been recovered from these beds (Orchard, 1978), which also contain species of the corals *Alveolites, Disphyllum, Favosites, Hexagonaria, Macgeea* and *Tabulophyllum*. At Western King Point, laminated crinoidal limestone fills hollows in the top of an underlying boundstone unit. Red, argillaceous, calcite mudstone and siltstone form interbeds and laminae with the limestone and also fill a complex fissure system (Braithwaite, 1967). The former have yielded conodonts, including *Palmatolepis triangularis, P. delicatula, P. subperlobata* and *I. alternatus,* indicative of the early Famennian, upper *triangularis* or lower *Palmatolepis crepida* zones (Orchard 1975, 1978).

The cliff section between Firestone Bay [4640 5350] and Eastern King Point [4669 5348] exposes stromatoporoid boundstone and rudstone together with crinoidal packstone and grainstone, comparable with those seen to the west. The boundstone contains undisrupted portions of crinoid stem up to 0.15 m in length [for example at 4656 5347]. In Durnford Street [465 536] where limestone and red calcareous siltstone and mudstone are interbedded, Orchard (1978) reported conodonts of the Frasnian *asymmetricus* Zone, including *Polygnathus asymmetricus unilabius, Palmatolepis* aff. *P. disparalvea* and *Ancyrodella africana.*

Near The Hoe, the lower part of the formation has been seen in boreholes in which thinly bedded limestone predominates. The middle and upper parts are present in West Hoe Park and at the waterfront cliffs and foreshore at The Hoe, where thicker beds are dominant. Crinoidal packstone with scattered stromatoporoids and corals grades imperceptibly into floatstone and rudstone, in which large colonial elements are preponderant. Boundstone units with tabular stromatoporoids and spar-filled shelter cavities are also common. Distinctive, thin beds of *Amphipora* bafflestone with a red, argillaceous wackestone matrix are also present. Orchard (1978) assigned massive limestone on The Hoe foreshore to the late Givetian and reported fragments of the conodonts *Polygnathus cristatus* and *Palmatolepis* cf. *disparilis* inferring an early Frasnian (lowermost *Polygnathus asymmetricus* Zone) age for strata in the West Hoe foreshore.

## CREMYLL AND EMPACOMBE

At Cremyll, on the west bank of the Hamoaze, limestone has been extensively quarried at the northern crop in the vicinity of the Ferry Terminal [453 534] and a southern crop occurs around the Mount Edgcumbe Park Orangery [455 532]. In the northern crop, the beds are inverted and dip steeply southwards; to the south, the limestone is the right way up and moderately to steeply inclined to the south-south-east. It is probable, although there is lack of intervening exposure, that the two crops are of the same sequence on opposing limbs of an asymmetrical antiformal fold.

These crops terminate westwards against the north-west-trending strike-slip Raveness Fault (see Figure 8), passing through Cremyll. In the north, the inverted limestone is juxtaposed across the fault with a similar limestone sequence, which is the right way up and moderately inclined south-eastwards. Cleaved, grey, silty mudstone and bedded basic tuff to the north appear to be subjacent.

The limestone is medium to thickly bedded, but may also be thinly bedded or massive. Fossils in the predominantly fine-grained, partially dolomitised wackestone and packstone (SDP 105, SDP 106) include stromatoporoid, coral, crinoid and shell debris. Brownish pink carbonate and silty mudstone occur as interlaminations, interbeds and wispy lenses throughout the sequence.

Near Empacombe, a limestone sequence, also southerly dipping and the right way up, is truncated by strike-slip faults and overlies cleaved, grey, silty mudstone on the foreshore [4425 5281]. This unit comprises beds (up to 1.5 m thick) of coarse, bioclastic limestone and thin interbeds of grey silty mudstone. Higher in the sequence, in Mount Edgcumbe Farm quarry [447 527], thin beds of medium grey, pelletoidal grainstone (SDP 110) are interbedded with paler, thickly bedded to massive, dolomitic packstone, with abundant crinoid and stromatoporoid fragments and pelloidal grains (SDP 111).

Conodonts indicative of the mid Frasnian *Polygnathus asymmetricus* Zone were recorded by Orchard (1978) from the foreshore by the Orangery. He also ascribed limestone in the Mount Edgcumbe Quarry to the Middle Devonian on the basis of conodont determinations.

## DRAKE'S ISLAND

Here, the Plymouth Limestone Formation dips and youngs southwards. The lowest part of the formation, thinly bedded micritic limestone, forms the northern foreshore. This limestone is in faulted contact to the south with thinly to very thickly bedded, basaltic tuff and hyaloclastite. Orchard (1978) recorded rafts of limestone within these volcaniclastic rocks which are succeeded conformably by some 10 m of limestone that is thinly bedded at its base and becomes massive upwards. This in turn is overlain by a thick basaltic lava and then by bedded tuffs, which extend southwards over much of the island. A few metres above the base of the tuffs there is a zone, several metres thick, of intense shearing that crosses the island parallel to the strike direction. This is interpreted as the location of a thrust that delimits the Jennycliff division of the Saltash Formation to the south.

Conodonts from the northern limestone and the bedded limestone within the volcanic rocks have been assigned to the mid Eifelian and Givetian respectively by Orchard (1978).

## MOUNT BATTEN

The strata on Mount Batten headland are moderately to steeply inclined southwards and the right way up. They are transected by north-west-trending faults, including splays of the Sutton Harbour Fault (Figures 1, 8), along which there are minor dextral displacements.

The lower part of the exposed succession, extending southwards from the northern foreshore, is a thick (about 150 m) sequence of thinly bedded limestone. The limestone is pale to bluish grey, very fine grained and commonly finely laminated. Locally, pink calcareous siltstone occurs as laminae and wisps within the limestone beds. A few thick beds in Castle Quarry [486 533] are richly macrofossiliferous. The succeeding predominantly volcanic sequence shows complex local variation where there is an interdigitation of lava, volcanic breccia, hyaloclastite and tuff, and grey silty mudstone with siltstone beds; near the base there is a packet of limestone beds. Above, and forming the top part of the formation, there is limestone which locally is thinly bedded, but elsewhere is intensely sheared, recrystallised (E 62789) and folded. This deformation is thought to result from thrusting at the top of the limestone and the base of overlying tuff of the Saltash Formation (compare with Chandler and McCall, 1985).

Lithostratigraphically, the sequence resembles that at Drake's Island and indeed, on the basis of the coral faunas, Taylor (1951) assigned the limestone at Mount Batten to the Givetian, although he thought that the younger rocks lay to the north in an inverted sequence. However, Orchard (1978) attributed the conodonts from the lowest part of the sequence to the mid Frasnian *Polygnathus asymmetricus* Zone.

## Depositional environment

The distribution of the lithofacies in the city charts the evolution and southward progradation of a mixed coral/stromatoporoid reef and of its peripheral facies. True reef facies within the formation appear to be confined to the adjacent Ivybridge district, notably in the Cattedown area, where reef development achieved its acme during the late Givetian (Orchard, 1977). Much of the Plymouth Limestone Formation in this district reflects the coeval accumulation of carbonate sediments peripheral to this reef.

Thin, lenticular, sharp-based, crinoidal packstones, present in the upper part of the Jennycliff division of the Saltash Formation, at Dunstone Point and Batten Bay, were deposited early in the Eifelian, and they provide evidence of the initiation of a carbonate build-up.

The Eifelian sequence of thinly interbedded, crinoidal packstone and mudstone in the Faraday Road Member marks the main transition from the earlier terrigenous regime. The thin, crinoidal packstone beds show the incursion of coarser, skeletal detritus into a low-energy, mud-accumulating environment. The erosional bases, normal grading and gradational tops indicate that emplacement was in response to short-lived, high-energy events. The packstone textures may reflect incomplete winnowing of fines and/or the subsequent infiltration of the carbonate mud matrix, possibly aided by bioturbation. The presence of fossiliferous lags composed of largely unabraded skeletal fragments does not reflect extensive or repeated transportation. The enveloping mudstone contains a fauna which suggests that much of the lag material comprises winnowed components of autochthonous benthic assemblages (Mayall, 1979). The stenohaline taxa in the enveloping mudstones relate to open-marine accumulation within the colonising depth range of contemporary shelly benthos. The intercalated hyaloclastitic basalt, basaltic tuff and tuffaceous mudstone are evidence of contemporary volcanism during the late Eifelian.

The thin- to medium-bedded, graded and locally cross-stratified packstone beds of the Prince Rock Member reflect the establishment, during the early Givetian, of a blanket shallow marine carbonate facies comparable to the lime sand belts recognised on many modern carbonate shelves and platforms. Deposition was probably below fair-weather wave-base, but well within the range of contemporary storm reworking. The associated beds of coral/stromatoporoid floatstone and boundstone provide the first evidence of prolific, in-situ growth and of sediment stabilisation by biohermal taxa. Rolled colonies in these beds record the limited effects of storm damage. In contrast, in the upper part of the member, rolled and abraded corals and stromatoporoids in rudstone and floatstone beds with channelled bases, represent transported debris, derived probably during major storms, from bioherms developing nearby. In the Cattedown area, and farther east, the small bioherms of this member were precursors to the main Givetian 'reef', which comprised a core facies of massive, coral/stromatoporoid boundstone and an interdigitating flanking facies of bedded, crinoidal packstone and floatstone, rich in reworked corals and stromatoporoids; both facies are included in the Cattedown Member of the formation.

The extensive cliff sections below The Hoe, and between Devil's Point and Eastern King Point, and along Richmond Walk, expose sequences of late Givetian and Frasnian strata contemporary with the Cattedown Member. These sections show the marked and complex, lateral and vertical transitions between crinoidal packstone/grainstone and tabular stromatoporoid floatstone and boundstone. The restricted faunal assemblage of the boundstone units indicates an environment disadvantageous to other contemporaneous taxa, but offering optimum conditions for tabular stromatoporoid colonisation. The success of such colonisation may have played a large part in excluding other faunal elements. Stromatoporoids with a tabular morphology were probably adapted to environments where low rates of sedimentation prevailed, but this has been thought to indicate either quiet-water conditions (Scrutton, 1977) or, as applicable here, a high-energy environment (Mayall, 1979). Here, deposition is thought to have occurred within a shifting patchwork of low-relief, stromatoporoid bioherms separated by spreads of crinoidal carbonate sand. Grainstone textures and local cross-bedding in the crinoidal deposits point to winnowing and tractional reworking in an open marine setting close to normal wave-base, whereas broken corals and stromatoporoids in graded rudstone beds are evidence of the damage and reworking caused by violent storms.

There is a significant east–west variation of late Givetian facies within the formation. Orchard (1975) has interpreted the western facies as a fore-reef deposit, but there is no evidence to suggest that a core reef facies, as seen at Cattedown, ever existed to the north. Mayall (1979) has suggested that a comparable facies contrast between the Brixham area and the Tor Bay reef complex was in response to greater subsidence in the former area, which inhibited reef growth (Scrutton, 1977). The close proximity of the facies in the Plymouth Limestone mitigates against differential subsidence as a mechanism here, but the facies contrast may reflect normal sedimentary processes consequent upon the formation and growth of the Cattedown reef and dependent on the local bathymetric conditions, which prevailed at the time. Clearly, it was the eastern part of the Plymouth Limestone build-up that was best suited to reef formation. This possibly lay on the windward side of the build-up, subject to the effects of wave and tidal action, and thus received the highest levels of nutrients. By inference, the western parts of the build-up were less exposed and less suited to core reef development and would have remained so, once the protective structure of reef was in place. However, the distinctive, mixed crinoidal and boundstone facies, which aggraded to the west of Cattedown, indicates open marine conditions; these possibly occurred on a submerged spit-like feature, which extended in a leeward direction from the eastern reef.

In contrast, the presence of wackestone and calcite mudstone, with a distinctive fauna of *Amphipora*, *Stringocephalus* brachiopods and turreted gastropods, and locally displaying desiccation fenestrae and cryptalgal lamination, reflects deposition in a truly restricted and locally emergent back-reef setting (Garland et al., 1996).

Such a facies occurs interbedded with, and succeeding, the core reef facies in the Cattedown area, but forms a subordinate element in sections in the city (for example at Devil's Point). The presence of at least 40 m of these shallow water sediments above the reef in Cattedown provides clear evidence not only for the southward (possibly south-eastward) migration of the main reef mass, but also of sustained aggradation, when reef growth was faster than the rise of sea level.

The fissures which provided Upper Devonian conodonts (Orchard, 1975, 1978) offer evidence of lithification, fracturing and foundering of the Plymouth Limestone build-up from mid Frasnian times onwards. These were associated with the termination of carbonate deposition in the area.

On Drake's Island to the south, the thinly bedded micrite assigned to the late Eifelian by Orchard (1978) appears similar to the Givetian (and possibly Frasnian) micritic limestone at Mount Batten. Braithwaite (1967) attributed those rocks to deposition in elevated, hypersaline, protected (back reef?) areas where import of bioclastic debris and other sediment was prevented. Supporting evidence is provided in the nearby Hooelake Quarry [496 530] by the abundance of *Amphipora ramosa* (Taylor, 1951), a specialised stromatoporoid thought to be indicative of restricted lagoonal/backreef environments (Scrutton, 1977). This may indicate the possible presence, to the south of the reef complex of Plymouth, of a second such system, the back reef deposits of which have been juxtaposed here by thrusting.

**Torpoint Formation**

The Torpoint Formation is newly named and this includes the purple and green facies rocks previously termed the 'Purple and Green Slates' in the west of the district (Burton and Tanner, 1986) and included in the Plympton Slate Formation of the Plymouth area by Chandler and McCall (1985). This formation interdigitates with the sedimentary and volcanic rocks of the Saltash division of the Saltash Formation. Although a complete section is lacking in this district, a variety of lithologies are accessible in river-side sections near Torpoint. The formation occurs mainly in the South Devon Basin, but it is also represented in the Tavy Basin and possibly in the Looe Basin. In the Looe Basin, it may occur thinly as the purple and green lithologies seen near Palmer Point [4417 5277] and Barn Pool [4552 5291].

Within the South Devon Basin, there are three informal geographical and stratigraphical subdivisions of the formation: the Plymouth, Skinham and Landrake divisions, named after the tracts in which they lie (Figure 20); a subordinate Central Plymouth 'subdivision' is also distinguished. The Torpoint Formation in the Tavy Basin is described after these.

Plymouth division

In the west, the division (Figure 20) occurs between the Cawsand and Portwrinkle faults, as two well-separated crops, in the north between Menheniot and Tideford Cross, and in the south between Bake and Port Eliot. In the Plymouth area, near Mutley [484 588], the main crop of this division is over 1 km wide and comprises predominantly reddish purple and subordinate green fine-grained sedimentary rocks, slaty mudstone and siltstone. Westwards through Devonport and the Torpoint peninsula, the Torpoint and Saltash formations interdigitate. The outcrop pattern hereabouts is also complicated by major early folds and faults, the latter including both early thrust faults and later extensional faults.

A lithologically distinct and structurally discrete *Central Plymouth subdivision*, which contains numerous beds of limestone, has been recognised in boreholes in central Plymouth (Figure 20a). The main crop is present eastwards from Millbay immediately to the north of that of the Plymouth Limestone Formation. It is also present beneath Millbay Docks [for example 468 541] between the northern and southern tracts of the Plymouth Limestone Formation. This sequence appears to represent an upward transition from rocks of the Saltash Formation that contain limestone with Givetian conodonts (see p.48). This subdivision may also include the purplish red cleaved mudstone of Sango Point and Sango Island [432 537], in which limestone nodules were noted by Ussher (1907).

Skinham division

This division (Figure 20) occurs in a tract between the Lynher Valley Fault Zone, near Notter [386 613] and the eastern margin of the district [489 587] near Manadon. It is present also in small fault-bounded blocks between Notter and Leigh Farm [338 641] to the north-west. The crop is up to 1.5 km wide in the Skinham area, between Saltmill Creek [429 598] and Kingsmill Lake [430 610], on the west bank of the River Tamar. The northern boundary is generally a thrust, which is flat lying to the east of the river. There is a thrust-faulted outlier [475 598] of the division to the north near Whitleigh. At the southern limit of the crop, there is a normal stratigraphical boundary to the east of the River Tamar but to the west the contact is faulted. Low-angle thrusts occur near Leigh Farm [339 644] and Wotton Farm [375 617].

Landrake division

In the Landrake area, between Trenance [346 619] and Trematon [395 595], this division occurs over a relatively small area (Figure 20). To the east, north and west of Landrake, the Torpoint Formation interdigitates with the Saltash Formation on a mappable scale. Hereabouts, tuff and lava from local volcanic centres interfinger complexly with both formations.

TORPOINT FORMATION OF THE TAVY BASIN

These rocks are present, with a faulted southern margin, from Leigh Farm [339 645], west of Pillaton to the eastern edge of the district (Figure 20). The northern boundary with the Tavy Formation is a stratigraphical transition. The faulted inlier near Vinegar Hill [400 639] also has a transitional boundary with the Tavy Formation.

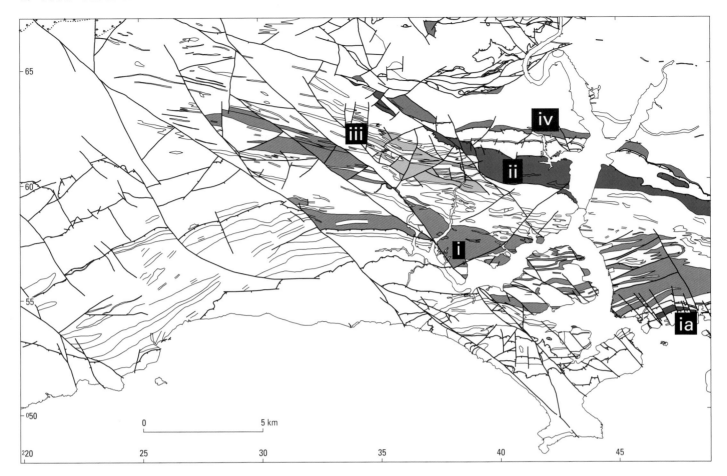

**Figure 20** Subdivisions of the Torpoint Formation.
i Plymouth division;   ia Central Plymouth subdivision;   ii Skinham division;   iii Landrake division;
iv Torpoint Formation of the Tavy Basin

There are also fault-bounded occurrences at Bealbury [371 668], near Halton Quay [406 656] and near Haye [422 668]. The first forms the main crop of the 'Bealbury Formation' of Whiteley (1983).

### LITHOLOGIES

Typically, the predominating cleaved mudstone and fine-grained siltstone are brownish purple to purplish red, whereas the coarse-grained siltstone and fine-grained sandstone are green, yellowish green or blue-green, although there are exceptions. The bedding is generally well defined by colour and markedly so in unweathered lithologies as seen in boreholes in Plymouth. Green reduction spots are sporadically developed in the mudstone. Colour mottling and colour transposition are present near some faults, joints and quartz veins, where there may have been fluid movement.

Interdigitation with grey rocks of the Saltash Formation has been mapped as a large-scale feature, and has been observed on much smaller scales down to those of laminae. On the north bank of the Lynher River, about Ince Point [405 565] and Shillingham Point [408 568], cyclic sequences of purple mudstone, green siltstone with sporadic sandstone beds, and grey, finely laminated mudstone and silty mudstone are developed.

The mudstone beds commonly appear to be structureless but some grade upwards from basal siltstone laminae. Siltstone is present as isolated laminae, some showing low-angle cross-lamination, as more regularly spaced laminae in sequences of green silty mudstone, and as more massive beds that have a diffuse parallel lamination. The sparse sandstone laminae and thin beds also generally show parallel lamination and some are normally graded.

In mudstone and siltstone at Bealbury [371 668], Whiteley (1983) recorded the presence of abundant siliceous nodules, up to 0.15 m across, that are richly fossiliferous.

### BIOSTRATIGRAPHY

In this district, Entomozoacean ostracods were recorded from purple and green slaty rocks by Ussher (1907). In purple and green rocks to the south of Warren Point [443 606], he also recorded the presence of ammonoids, that House (1963) attributed to the mid Frasnian *Manticoceras cordatum* Zone. A biostratigraphical summary of the slate near Plymouth (Gooday, 1973, 1974) focused on the ostracods, but tentaculitids and *Styliolina* were also documented. Gooday's distribution pattern of ostracods with Frasnian and Famennian ages matches with the crops of this formation.

During this resurvey, new collections of ostracods (Plate 8) were made and the results of analyses by Wilkinson were recorded (see Information sources). The ages obtained accord with the observed structure of both the Plymouth and Skinham divisions. In the Plymouth division (Wilkinson, 1987b, 1990a), on the west bank of the River Tamar, between Sand Acre Bay [419 574] and the Lynher River, where near-vertical bedding youngs northwards, the faunas show a transition from the mid/late Frasnian (for example *Entomozoe [Nehdentomis] pseudorichte-rina*) to Famennian (for example *Richteria [Nehdentomis] ser-ratostriata*). South of the Lynher River, the same Frasnian–Famennian transition is present by Thankes Lake [436 556] (Wilkinson, 1990a) and on the south bank of the Lynher River [412 563], west of Antony House (Wilkinson, 1990a, 1990b). This sequence is involved in asymmetrical, northward-verging, large-scale folds, and youngs southwards. Late Famennian palynomorphs have been obtained from the Saltash Formation, where it interdigitates with the Torpoint Formation on the southern foreshore at Torpoint (McNestry, 1994a).

In the Skinham division, the ostracods show a similar range of ages. Early to late Famennian ages (Wilkinson, 1987b) were obtained from faunas from near to the northern, thrust-faulted boundary in the Tamar valley, where the bedding is steeply inclined to overturned northwards. To the south of Skinham Point [431 607], where the bedding is gently inclined, mid and late Frasnian ostracods were recorded (Wilkinson, 1987b, 1990b, 1993a). Ammonoids identified as *Manticoceras* cf. *cordatum* (Wilkinson, 1987a, 1993a), the index fossil for the mid Frasnian, have been collected from localities some 200 m south of Skinham Point and some 100 m south of Warren Point [444 604]. Upper Devonian ostracods have been recorded from elsewhere in these two divisions and from the Landrake division.

Within the Tavy Basin, ostracods from Vinegar Hill indicate a Frasnian age (Gooday, 1973), whereas the southernmost crop has yielded ostracods, including *Maternella hemisphaerica*, *Richterina {Richterina} striatula*, *Richterina {Richterina} costata*, of late Famennian age (Wilkinson, 1990a). The conodonts, ostracods, ammonoids and trilobites obtained by Whiteley (1983) from his 'Bealbury Formation' also provided a late Famennian age.

DEPOSITIONAL ENVIRONMENT

The predominantly fine-grained nature of these sediments indicates a distal or shielded environment. The coarser deposits have features consistent with deposition from turbid flow. The grading and other sedimentary structures are typical of distal turbidites. The thicker laminated beds are either composite beds or represent deposits of larger suspended turbid clouds. It is probable that at least some of the mudstone beds, with thin, coarser bases may also represent turbid flow deposition. Although bioturbation is not uncommon, there is a notable absence of benthonic faunas but a presence of ostracods, ammonoids and styliolinid gastropods. This possibly indicates disaerobic bottom conditions in an open marine environment.

The Entomozoacean ostracods are planktonic and nektoplanktonic species not known from neritic facies (Wilkinson, 1993b) and are considered to have accumulated in an outer shelf setting (Selwood et al., 1984) or with rise deposits (Stewart, 1981b; Selwood et al., 1984). Their presence typifies the German *Cypridienschiefer* basinal facies of Schmidt (1926), in which similar ostracod-rich lithologies are associated with posidoniid bivalves and ammonoids. Bandel and Becker (1975) estimated water depths of 500 m to 1000 m for this facies in the southern Rhineland.

GEOCHEMISTRY AND PROVENANCE OF THE TORPOINT FORMATION

The geochemical data for the purple and green rocks of the Torpoint Formation and the grey rocks of the Saltash Formation, plotted on a $Zr/TiO_2$ v. $Nb/Y$ discriminant diagram (Winchester and Floyd, 1977), occupy a very restricted field, equivalent to andesitic and basaltic rocks. This contrasts markedly with the subalkaline to alkaline nature of the volcanic and hypabyssal rocks present in this district (Figure 26). The $Zr/TiO_2$ data for the slaty mudstone indicates a more evolved crustal source for the sediments and a sparse contribution of volcanogenic sediment to the South Devon Basin. The petrography and geochemistry of the variegated rocks do not reveal major discriminating differences. The purplish red mudstone is rich in oxidised (trivalent) iron, and contains low concentrations of copper compared with the grey lithologies, which also have significant concentrations of nickel (>100 ppm). Both the red and grey rocks are depleted in rubidium, barium, cerium and zirconium relative to average shale.

Ziegler and McKerrow (1975) demonstrated that a red colouration in non-abyssal sediments is not an indicator of nonmarine or shallow water environments, and that red and grey-green colours were not necessarily simply a function of oxidising (high Eh/pH) or reducing (low Eh/pH) conditions above or below a water table (compare with Thompson, 1970; McBride, 1974). The roles of supply of oxidised detritus (with ferric iron), of rate of burial, and of reduction, in the presence of organic matter, were identified as important for the preservation of red rocks in an offshore marine setting. The analogous variegated sediments in the Rhenisches Schiefergebirge of Germany were interpreted by Franke and Paul (1980) as continentally derived sediments. They confirmed that the red colouration is due to ferric iron, developed during continental weathering and inferred transport as hydroxide floccules or coatings on clay minerals. The oxidised state was maintained during fluvial discharge and sedimentation took place in deeper environments from gravity flows. They thought that dilution of pigment caused the absence of red colouration in the coarser sediments. The presence of green fine-grained sediments was ascribed to the reduction of ferric iron during the diagenetic oxidation of organic matter.

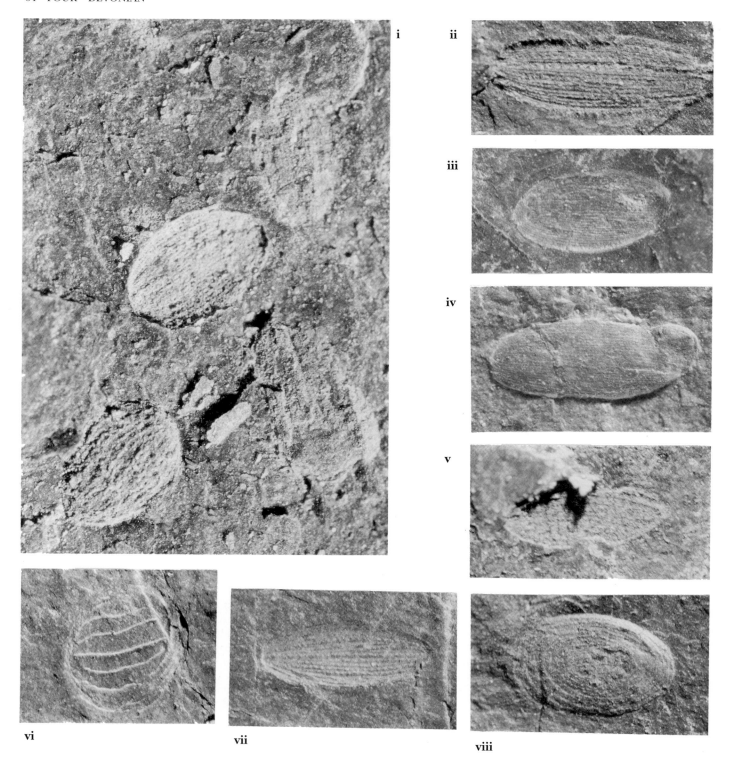

Regionally, the purple and green Upper Devonian sedimentary rocks in south-west England also include the fluvial and shallow marine rocks of north Devon and Somerset (Edmonds et al., 1975), and discrete sequences interbedded with the flysch of the Falmouth Series of Hill and MacAlister (1906) and Leveridge et al. (1990).

The maintenance of colour integrity in differing environments, fed by transport systems of varying extent, indicates that organic content alone is not the determinant of red or green colouration and may point possibly to different sources for the sediments. In this district, the purple and green strata are seen to have major

**Plate 8** Late Frasnian to late Famennian Entomozoacean Ostracoda (Crustacea) from the Torpoint Formation (GS516). Tectonic distortion is illustrated by comparing figures of *Richterina costata* (i, ii and vii) and *Richterina striatula* (iii and iv). All specimens, except *Nehdentomis pseudorichterina*, are from the *Maternella hemisphaerica-dichotoma* Zone. (Magnification: 1 × 50, all others × 20). Specimens are in the Biostratigraphy Collections of BGS, Keyworth.

i. A cluster of *Richterina costata* (Richter, 1869) from a cliff 25 m to the east of Tamerton Bridge [4475 6051]. NB221

ii. *Richterina costata* (Richter, 1869) from Crownhill Fort, north Plymouth [4868 5935]. NB415a

iii. *Richterina striatula* (Richter, 1848) from Crownhill Fort, north Plymouth [4868 5935]. NB417

iv. *Richterina striatula* (Richter, 1848) from Crownhill Fort, north Plymouth [4868 5935]. NB406a

v. *Nehdentomis pseudorichterina* (Matern, 1929) from the Saltash bypass/A388 junction [4136 5935]. NB274. Late Frasnian.

vi. *Maternella hemisphaerica* (Richter, 1848) from Crownhill Fort, north Plymouth [4868 5935]. NB406b

vii. *Richterina costata* (Richter, 1869) from Crownhill Fort, north Plymouth [4868 5935]. NB413

viii. *Maternella dichotoma* (Paeckelmann, 1913) from Crownhill Fort, north Plymouth [4968 5935]. BN415b

developments to the east that thin and die out westwards. This probably indicates that the emplacement of this sediment was from the east.

**Tavy Formation**

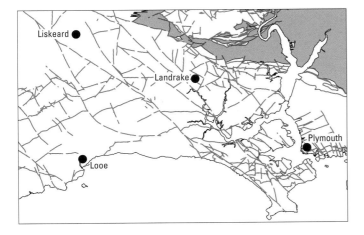

This newly defined formation occupies much of the north-eastern part of the district and its crop falls within an area of rocks designated as Upper Devonian by Ussher (1907). It forms the main stratigraphical component of the Tavy Basin and it is mainly exposed along the banks of the River Tavy. This green slaty mudstone is replicated elsewhere in the region: the formation compares closely with the Tredorn Slate Formation of north Cornwall (Selwood et al., 1998),

formerly known as Tredorn Slate, Delabole Slate and the Woolgarden Phyllites, and with the Kate Brook Slate of the Newton Abbot area (Waters, 1974).

The southern margin of the formation is mainly a stratigraphical boundary with the Torpoint Formation, as at Weir Point [4390 6185], but in the west it is delimited by elements of the Lynher Valley Fault Zone and the Ludcott Fault. The upper boundary with the Burraton Formation is transitional, to the west near St Ive Cross [316 672] and in the Tamar valley, in the extreme north of the district. In the latter area, the succession appears to be inverted, as in the vicinities of North Hooe Mine [4254 6608] and Hole's Hole [425 655]. The boundaries with the Carboniferous rocks of the St Mellion Outlier are faulted.

The formation is predominantly a pale green to greyish green, lustrous, well-cleaved, chloritic, silty mudstone. In places, the mudstone shows faint colour banding, and the staining of joints and cleavage planes by iron and manganese is common. Pyrite is ubiquitous, in cubes up to 50 mm across. The coarser lithologies include laminated siltstone and fine-grained sandstone. The sandstone may also occur as lenses and sporadic, sinuous burrow-fills. This lithology is locally more prominent, as at Hole's Hole [4302 6531] where fine- to coarse-grained beds, up to 0.1 m thick, are interbedded with mudstone in a sequence that is 3 to 4 m thick. These sandstone beds have sharp bases and tops, and show low-angle cross-lamination.

The main sandstone in this formation, the **Trehills Sandstone Member**, occurs between Thorn Point [440 629] and Porsham Farm [487 620] in the southern part of the crop and it has a maximum thickness of 50 m. It comprises fine- to medium-grained sandstone beds, up to 1.2 m thick, interbedded with mudstone and laminated micaceous siltstone. The sandstone beds are sheets with sharp erosional bases, tops that are sharp or show delayed grading, and commonly show parallel or ripple cross-lamination. This member comprises litharenites of predominantly monocrystalline quartz, subordinate polycrystalline quartz, detrital muscovite and degraded igneous rock fragments, in a matrix of authigenic illite and goethite, locally replaced by chlorite (E 72448/9).

There are sporadic minor occurrences of basic volcanic rocks within the formation, principally vesicular lava, which occur near Warleigh House [4567 6171], on the bank of the River Tamar near Lockridge Farm [4319 6637] and in the railway cutting [4371 6692] west of Bere Alston.

BIOSTRATIGRAPHY

The Tavy Formation is sparsely fossiliferous and Ussher's (1907) attribution of these rocks to the Upper Devonian was based only on the occurrence of the ostracod *Entomis*. Poorly preserved, non-diagnostic faunas consisting of bivalves, brachiopods, crinoid debris and indeterminate ostracods were reported from several localities in the district by Whiteley (1983) but his assignment of these rocks to the late Famennian *Maternella hemisphaerica–dichotoma* ostracod Zone and the *Bispathodus costatus* conodont Zone was based on fossil evidence from adjacent areas to the north.

During the recent survey, only a sparse macrofauna, comprising *Cyrtospirifer* and crinoidal debris, was observed. Wilkinson (1987b) identified a single entomozoid ostracod *?Maternella* sp. from a section [4710 6398] on the River Tavy and suggested that a late Famennian age was likely. Samples from other localities containing *Tentaculites* and *Styliolina* indicate an age no younger than late Frasnian (Wilkinson, 1987b). A variety of spores and acritarchs obtained from slate along the River Tamar [4258 6613], near the northern edge of the district, provided strong evidence of a late Famennian age (Molyneux and Owens, 1990). Spores and acritarchs from a section [4683 6362] on the River Tavy indicate a late Frasnian to Famennian age, but they are not definitive (Dean, 1992). Stewart (1981a) concluded that this formation ranged in age from the upper *Palmatolepis marginifera* Zone to the topmost part of the Devonian, but a late Frasnian to late Famennian range is more probable.

DEPOSITIONAL ENVIRONMENT

The presence of marine fossils in a predominantly argillaceous sequence might suggest deposition in quiet-water conditions, below storm wave base, in which muds settled from suspension. However, the lithological uniformity of major parts of the formation may indicate rapid sedimentation from turbid flow. The ubiquitous presence of chlorite and pyrite indicates the significant role of reducing conditions during and/or after deposition. Whether or not this material reflects an original organic content, the sourcing and depositional regime certainly differed from the more variable Torpoint Formation. The sheet sandstone beds of the Trehills Sandstone Member were interpreted (Jones, 1993) as turbidites, the cross-lamination being indicative of deposition from low energy unidirectional, tractional flows. Their repetitious occurrence indicates that they may represent part of a fan lobe sequence. The wave ripples on the tops of the turbidites at Hole's Hole were taken by Jones (1993) to indicate probable reworking below a fluctuating wave base in shallow water.

## Burraton Formation

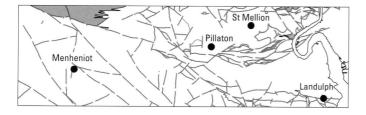

The crop of the Burraton Formation hereabouts corresponds to that of Whiteley (1983) who defined the formation from exposures around East Burraton [4135 6730] immediately to the north of this district. Whiteley considered that this formation was equivalent to the California Quarry Formation of Stewart (1981a) in the Tavistock and Boscastle districts. It also bears strong similarities to the Yeolmbridge Formation (Stewart, 1981b) and the South Brentor Formation (Isaac, 1983) of the Tavistock district, although these formations are somewhat more calcareous. The Hyner Shale and Trusham Shale (Selwood et al., 1984) of the Teign valley, on the eastern side of Dartmoor, were also considered to be comparable to the Burraton Formation by Whiteley (1983).

Within this district, the Burraton Formation has a fragmentary crop. In occurs most extensively along the northern boundary of the district, west of Ludcott Farm [303 664]. Here, the eastern boundary is one of interdigitation with the rocks of the Tavy Formation. To the west of the River Tamar in the Halton Quay area [412 653] and along the northern boundary of the district in the Haye [421 669] area, the Burraton Formation displays a gradational contact with structurally overlying rocks of the Tavy Formation. In this area, however, the structurally lower boundary of the Burraton Formation is probably a thrust. At Fursdon [410 669], the Burraton Formation has a gently inclined boundary with the Tavy Formation, and approximately 1 km to the east on the western bank of the River Tamar, a nappe of Tavy and Burraton formations overthrusts the Carboniferous St Mellion Formation. The position of this locality immediately across the river from the North Hooe Mine, where the Burraton Formation appears to rest on the Newton Chert Formation, suggests that the lower boundary of the Burraton Formation at North Hooe is also tectonic.

On the eastern bank of the River Tamar, the formation has isolated exposures at North Hooe Mine [425 661] and Hewton [426 655], where it structurally underlies, with a gradational contact, rocks of the Tavy Formation. Although a lower boundary is not seen in these outcrops, black sooty slaty mudstone debris on the North Hooe Mine dumps may indicate that the Burraton Formation overlies rocks of the Newton Chert Formation at depth.

The Burraton Formation comprises dark grey to black, lustrous, slaty mudstone with grey siltstone laminae, wavy and cross-bedded in part. Subordinate grey-green, thin to thick, medium- to coarse-grained sandstone beds also occur. Exposures in the district show a general sequence in ascending order of i) dark grey and dark blue-grey, lustrous, slaty mudstone, weathering grey-green, with dark grey, micaceous, slaty mudstone in places; ii) black, siliceous, slaty, silty mudstone and dark grey, slaty, silty mudstone with sporadic grey silty laminae iii) laminated slaty mudstone. The sporadic, purple, buff and green, slaty mudstone beds reported within this formation by Whiteley (1983) have not been recorded in the district. The true thickness of the Burraton Formation is not known, although Whiteley (1983) estimated its structural thickness as approximately 100 m.

BIOSTRATIGRAPHY

The Burraton Formation of this district has yielded poorly preserved ammonoids and ostracods from nodules within the grey slate at Burraton [4134 6710]: they include *Kosmoclymenia, Costaclymenis* cf. *enodis, Imitoceras, Richterina*

*(R.) costata* together with a hindeodellid conodont (Whiteley, 1983). This author suggested a late Famennian age (Clymenia Stufe). Miospores from Hewton [426 655] were identified by McNestry (1993) as *Auroraspora* sp., *Diducites versabilis, D.* spp., *D. mucronatus, Dibolisporites* sp., and *Grandispora* cf. *cornuta.* These, together with the acritarch *Maranhites* sp. indicate that the formation ranges up to the top of the Famennian (Wocklumaria Stufe).

DEPOSITIONAL ENVIRONMENT

The mud-dominated parts of the Burraton Formation are interpreted to represent deposition from suspension in a low-energy regime in a marine environment. Elsewhere, the presence of cross-bedded, arenaceous laminae is indicative of a relatively high-energy regime, possibly by turbid flow on a shelf or in a basinal area subjected to bottom currents.

# FIVE

# Carboniferous

The presence of Carboniferous rocks within the district was first recorded by Holl (1868), who noted 'outlying patches of Culm' around St Mellion and later by Hinde and Fox (1895, 1896) and Fox (1896) who reported radiolarian-rich rocks of 'Lower Culm' age at St Dominick, Pillaton and Paynter's Cross. As a result of the first systematic geological survey of this district, Ussher (1907) divided these rocks into a sequence of shales with radiolarian cherts of 'Lower Culm' age and a greywacke sequence of 'Middle Culm' age. Although Ussher (1907) described the relationships between the Upper Culm, Middle Culm and underlying Upper Devonian slate as enigmatic, he rejected the possibility that they were emplaced along horizontal thrusts and suggested instead that the sequences were separated by unconformities. The opposite view was taken by Whiteley (1983) who described the Carboniferous rocks, that form an outlier at St Mellion, as a stack of three thrust nappes. He subdivided these strata into a number of formations characterised by the presence of sandstone, mudstone or chert (see Table 11). This resurvey of the district has resolved Whiteley's divisions into three mappable lithostratigraphical units: the Brendon, St Mellion and Newton Chert formations (Table 11). These suites of sediments represent a late stage in the evolution of basins in this district, synchronous with the onset of inversion and deformation of the more southerly basins of the passive margin (see Chapter 11).

## Brendon Formation

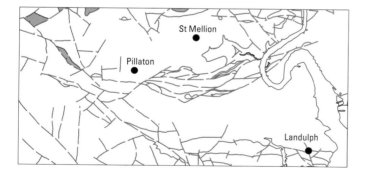

The Brendon Formation, as redefined here, broadly equates with that of Whiteley (1983), apart from the exclusion of his Newton Chert Member which has now been assigned formational status. Within this district, this formation is restricted to small isolated crops within the St Mellion outlier, principally along its western side. Its relationships with the other units of the outlier and with the underlying Tavy Formation appear to be exclusively tectonic. The apparent stratigraphical, subhorizontal

contacts between the Brendon and Newton Chert subdivisions led Whiteley (1983) to suggest that the Newton Chert formed a member within the Brendon Formation. Those contacts are now mapped as thrusts, as in fact were Whiteley's boundaries between the Brendon Formation, his Carboniferous Crocadon Sandstone and Cothele Sandstone formations, and the underlying Devonian rocks, in the St Dominick area to the north of the district. Because the Brendon Formation of the St Mellion outlier appears to be entirely allochthonous, forming a series of subhorizontal thrust nappes, little is known of its thickness, stratigraphical position, or its relative setting within the sedimentary basin. Whiteley (1983) estimated its minimum structural thickness as 450 m but it could be significantly thicker.

This formation comprises a sequence of dark grey, locally siliceous, slaty mudstone and siltstone, with thickly laminated to thinly bedded siltstone. Sporadic, blue-grey to grey-green, coarse-grained greywacke sandstone forms dispersed packets of thin graded beds, with dark grey mudstone. To the north of the district, the Brendon Formation commonly contains dolerite and picrite intrusions, vesicular lavas and dispersed thin tuff beds (Whiteley, 1981, 1983). Examples close to the margin of the district include doleritic intrusions near Greenswell [3588 6741] and Tipwell [3857 6727] and a picrite at Park Farm [3502 6700].

### BIOSTRATIGRAPHY

Rare faunas, including the conodonts *Polygnathus communis communis*, *Hindeodella* cf. *ibergensis* and *H. undata*, together with fragments of ammonoids, bivalves and podocopid ostracods, have been recorded from the Brendon Formation outside this district (Whiteley, 1981). However, no zonal taxa have been recovered and the age of the formation can only be defined as Viséan in part. On the basis of the presumed stratigraphical continuity with his overlying Newton Chert Member, Whiteley, (1983) suggested that the lower age limit of the formation might be Tournaisian. The presence of bivalve-bearing '*Posidonia (Lamellibranch)* Beds' within both the Brendon Formation and the Newton Chert Formation (Whiteley, 1983) suggests, however, that these two formations are in part contemporaneous.

### DEPOSITIONAL ENVIRONMENT

The evidence for the sedimentological environment of the Brendon Formation is sparse. The fossil evidence indicates marine conditions, whereas the characteristics of the sandstone beds which include comminuted plant debris, may reflect relatively distal turbidite emplacement. Whiteley (1983), using the *Becken and Schwellen* facies model of Schmidt (1926), characterised the Brendon Formation, together with its correlatives (the

**Table 11** Correlation between the units in the St Mellion outlier according to Whiteley (1983) and those mapped in this survey.

| Whiteley, 1983 | | This memoir | |
|---|---|---|---|
| Newton Chert Member (Brendon Formation) | late Viséan | Newton Chert Formation | Viséan |
| Pillaton Chert Formation | ? late Viséan | | |
| Chert olistoliths in Crocadon Sandstone Formation | early Viséan | | |
| Crocadon Sandstone Formation | ? Namurian to Tournaisian | St. Mellion Formation | Namurian to Tournaisian |
| Cothele Sandstone Formation | ? Tournaisian | | |
| Brendon Formation | Namurian to ? Tournaisian | Brendon Formation | Viséan and ? Tournaisian |

Upton Wood Formation of Stewart (1981a), the Greystone Formation of Turner (1982) and the Lydford Formation of Isaac (1983), as 'Restricted Basin Facies', although the local palaeogeography was not described.

## St Mellion Formation

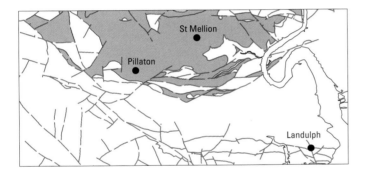

Whiteley (1983) divided the Carboniferous greywacke rocks of the St Mellion outlier into two contemporaneous lithostratigraphical units; the Cothele Sandstone Formation and the Crocadon Sandstone Formation. The former was distinguished by the presence of sporadic thicker beds of sandstone (up to 2 m thick) and a relatively high proportion of mudstone interbeds, whereas graded bedding and sole structures were noted to be more common in the Crocadon Sandstone Formation. However, this resurvey has shown that there is considerable variation within both of Whiteley's units, such that they cannot be distinguished. These two units are here combined, forming the St Mellion Formation.

This formation is the major component of the St Mellion outlier and has a large crop, which is entirely bounded by thrusts and other faults, on the west side of the River Tamar to the north of Pillaton [365 643]. It appears to occur in one major and several minor thrust nappes within the St Mellion outlier. Because of this tectonic isolation, the stratigraphical relationships and the palaeogeographical setting of the formation are not known. The contact with the underlying Devonian rocks

of the Tavy and Burraton formations is a major horizontal thrust along which lenses of chert (Newton Chert Formation) and, less commonly, of purple slaty mudstone (Torpoint Formation) are included as tectonic 'horses'. The full thickness of the St Mellion Formation is unknown although geophysical evidence suggests that the structural thickness of the major exposed thrust nappe formed by the formation is of the order of 750 m.

The St Mellion Formation comprises interbedded sandstone and mudstone with sporadic siltstone beds (Plates 9, 10). The sandstone tends to be localised within discrete units which also contain subordinate interbedded mudstone; isolated beds of siltstone and sandstone also occur elsewhere. Rare beds of polymict, conglomeratic sandstone are also present.

Where sandstone dominates the succession, it is fine to medium grained and almost black when fresh, but commonly weathered to blue-grey, grey-green or buff. Typically, these sandstone beds are tabular, 0.2 m to 0.6 m thick, and are regularly interbedded with thin, dark grey mudstone, but thicker single beds and composite beds, 2 m to 5 m thick, are present locally (Figure 21). The sandstone beds commonly show normal grading, sharp bases with sole structures, and in places rippled tops (Figure 22). The thin, fine-grained sandstone beds which occur within the thicker mudstone sequences may display sedimentary lineations on the bedding surfaces or parallel lamination. Locally, both thin and thick beds of sandstone may show an absence of internal sedimentary structures, apart from amalgamated bedding (Jones, 1993).

The sandstone is variable between feldspathic litharenite and sublitharenite, containing angular and subrounded grains of both unstrained and strained, monocrystalline and polycrystalline quartz, with significant amounts of potassium feldspar and detrital mica (Plate 11), as well as albite (Whiteley, 1983). The lithic fragments include chert, mudstone, metamorphic, and igneous clasts, with rare grains of zircon and tourmaline (Jones, 1993). Conglomeratic sandstone occurs at the base of a few sandstone beds and this comprises rounded pebbles of quartz with subordinate pebbles of chert, mudstone and sandstone.

Isolated beds of very poorly sorted sandstone with abundant plant debris are sporadically present as

**Plate 9** Sheet bedded sandstone of the St Mellion Formation, quarry north-west of Heathfield Farm [3945 6655]. Quarry face, viewed northwards, is approximately 4 m high (GA517).

**Plate 10** Thickening- and coarsening-up inverted sandstone sequence in the St Mellion Formation at Paynter's Cross road cutting [3998 6396]. Looking eastwards (GS518). Hammer is 0.30 m long.

distinctive lithologies. They are dark grey, muddy, markedly micaceous and fine- to medium-grained. They display irregular wavy bedding surfaces and a lensoid form, in contrast to the tabular bedding of most other sandstones within the formation. Some slump structures are noted locally. The sandstone is composed of angular grains of monocrystalline and polycrystalline strained quartz, supported in a matrix of micaceous, muddy siltstone. The plant debris comprises predominantly plant stems of *Cordaites* type, up to 0.2 m in length.

The sequences of dominantly mudstone exceed 30 m in thickness and comprise thick beds of dark grey mudstone and silty mudstone. In these, siltstone occurs as laminae in packets up to 1.5 m in thickness. The laminae exhibit parallel lamination, low-angle cross-lamination and climbing ripple cross-lamination.

### BIOSTRATIGRAPHY

Apart from abundant plant fragments, few fossils have been reported from the St Mellion Formation in this district. Those recorded include ammonoids, ostracods, conodonts and plant spores and they indicate that the formation ranges in age from at least the lower part of the Tournaisian to the upper part of the Viséan (Table 12). Conodonts from Mount Pleasant [3907 6692], including *Siphonodella cooperi, S. duplicata, S. quadruplicata, Polygnathus communis communis, P. triangulus, Pseudopolygnathus radinus* and *Elictognathus laceratus,* were interpreted by Whiteley (1983) as representative of a Tournaisian (*Siphonodella sandbergii* zone) age. Goniatites of the genera *Ammonellipsites, Gattendorfia* and *Muensteroceras,* from a road section at Jubilee Cottage [3980 6655], have been reported by Matthews (1970) to indicate a mid to late Tournaisian age. Re-examination of the ammonoids by Dr N J Riley (personal communications, 1997, 2001) has indicated that they are of Namurian age. They belong to one of the

*Nuculoceras nuculum* (E2c2-4) marine bands of the latest Arnsbergian (early Namurian). *Nuculoceras nuculum* (*Ammonellipsites* of Matthews), *Eumorphoceras* sp. juv (*Gattendorfia* of Matthews), *Zephyroceras darwenense* (*Gattendorfia* of Matthews) and Beyrichoceratoides (*Muensteroceras* of Matthews) are present. This is the first record of the *N. nuculum* assemblage in the province, and Riley reports that it provides a fix for the Global Mid-Carboniferous boundary in the local sequence. A similar fauna in a nearby quarry at Heathfield [3974 6610], reported by Whiteley (1983), included *Ammonellipsites princeps* and *Muensteroceras complanatum* identified by Matthews. The late Tournaisian age attribution by Whiteley is thus in doubt. An impoverished spore flora, including *Lycospora pusilla, Tripartites* and *Punctatisporites,* is present within the St Mellion Formation and this indicates that the age ranges up to the late Viséan and possibly into the early Namurian (Whiteley, 1983).

### DEPOSITIONAL ENVIRONMENT

The fauna indicates that the formation was deposited in a marine environment. The sandstone beds appear to represent unconfined turbidite flows with no evidence of wave action. This suggests that the formation was deposited below wave base, although the exact depth of water cannot be determined. This conclusion may conflict with Whiteley's (1983) report of a rootlet bed in a temporary excavation to the north of the district.

The sedimentological evidence indicates that the St Mellion Formation represents a prodelta sequence linked to coastal deltaic systems (Jones, 1993). Individual sandstone beds were deposited during periods of high discharge in the deltaic feeder system when turbiditic flows passed down the delta front into the sedimentary basin. The packets of turbiditic sandstone beds represent more prolonged episodes of delta progradation into the moderately deep marine sedimentary basin. Periods of instability

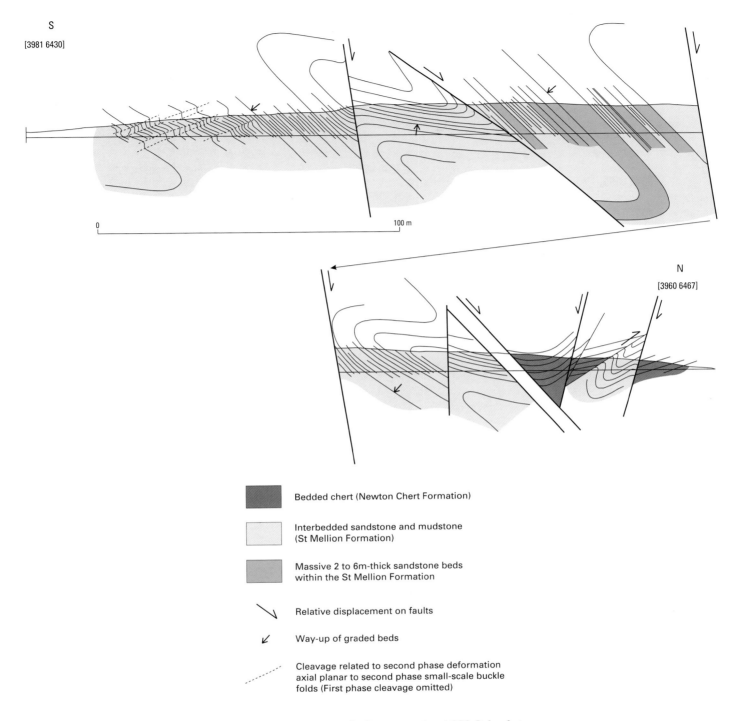

Bedded chert (Newton Chert Formation)

Interbedded sandstone and mudstone
(St Mellion Formation)

Massive 2 to 6m-thick sandstone beds
within the St Mellion Formation

Relative displacement on faults

Way-up of graded beds

Cleavage related to second phase deformation
axial planar to second phase small-scale buckle
folds (First phase cleavage omitted)

**Figure 21**  Section through the road cutting at Paynter's Cross, on the A388 Saltash to
Callington Road, showing the general inversion of the St Mellion Formation in large-scale
southward-facing first phase folds. Note the massive sandstone beds within the more thinly
interbedded sandstone and mudstone sequence of the St Mellion Formation.

**Figure 22**  Graphic logs of the St Mellion Formation.

a.  Lower face of quarry north-west of Heathfield Farm
b.  Upper face of quarry north-west of Heathfield Farm
c.  Quarry south of Heathfield Farm

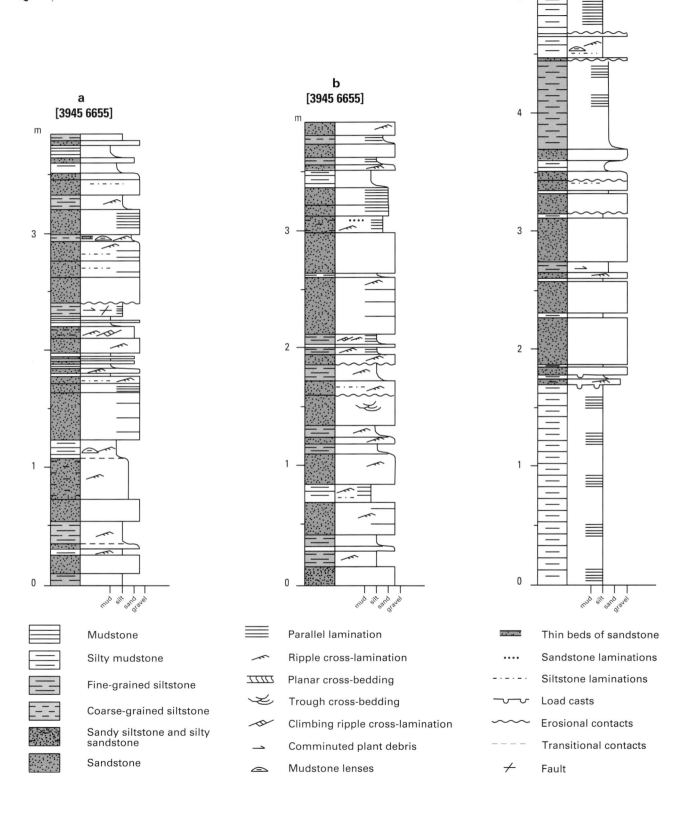

| | |
|---|---|
| Mudstone | Parallel lamination |
| Silty mudstone | Ripple cross-lamination |
| Fine-grained siltstone | Planar cross-bedding |
| Coarse-grained siltstone | Trough cross-bedding |
| Sandy siltstone and silty sandstone | Climbing ripple cross-lamination |
| Sandstone | Comminuted plant debris |
| | Mudstone lenses |

| |
|---|
| Thin beds of sandstone |
| Sandstone laminations |
| Siltstone laminations |
| Load casts |
| Erosional contacts |
| Transitional contacts |
| Fault |

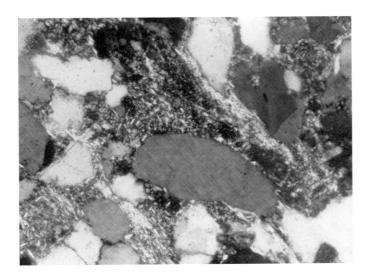

**Plate 11** Photomicrograph of a thin-section of sandstone from Heathfield Quarry. Monocrystalline quartz, with overgrowths, dominant, quartz and illite lithic grain bottom left. Quartz and illite form pseudomatrix throughout (GS519) × 154.

within the accumulating pile of sediment are marked by the deposition of slumped plant-rich beds derived from the delta front. The preservation of recognisable plant stems, up to 0.2 m long, within these beds suggests that transport within the sedimentary basin was limited.

The presence of common plant material demonstrates a continental source for the sediment. This source area comprised deformed low-grade metamorphic rocks which provided most of the component grains of the sandstone beds. The significant proportion of potassium feldspar also indicates the presence of exposed granitic rocks within the source area.

## Newton Chert Formation

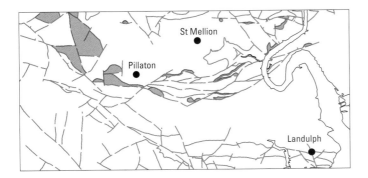

The Newton Chert Formation as described here embraces the following, nominated by Whiteley (1983): the Newton Chert Member of the Brendon Formation, the Pillaton Chert Formation, and the fragmentary chert bodies, of the Crocadon Sandstone Formation, which he thought to be olistoliths. The formation is of a similar age and lithology to the Teign Valley Chert (see Whiteley, 1983), the Fire Beacon Chert (Freshney et al., 1972) and the Meldon Chert (Edmonds et al., 1968) which may be correlatives.

This formation has a fragmentary distribution within the St Mellion outlier. In the western part of the outlier, it is thicker and less tectonised than elsewhere, occurring in faulted blocks at Herod Down [356 646], Tor Wood [358 650], Crendle Down [342 655], Newton Ferrers [347 660] and Durnaford [334 669]. In a few places along the southern boundary of the outlier, bodies of Newton Chert, a few metres in length, occur as slivers within normal fault zones. In the eastern part of the outlier, the formation forms a series of generally thin, highly disturbed lenses lying along subhorizontal thrusts at the base of the major nappe of the St Mellion Formation, and also along smaller thrust slices of the same formation. Because all of its boundaries are tectonic, the stratigraphical relationship and environmental setting of the formation are not known. However, the presence of 'Posidonia Beds' within both the Newton Chert and within the Brendon Formation (Whiteley, 1983) indicates that they are in part contemporaneous. The full thickness of the Newton Chert Formation is unknown but an estimated structural thickness is about 100 m: this is speculative because of the degree of internal disharmonic folding and thrusting.

The lower part of the formation is a sequence of massive, black or dark grey, siliceous, slaty mudstone and siltstone, which structurally underlie the main development of chert that is characteristic of the formation. In the western part of the St Mellion outlier, the siliceous slate is interbedded with thin beds of very dark grey, sooty shale and tabular beds of glassy, black chert. At Halton Quay [4166 6573] to the east, the mudstone encloses rare lenses of fine-grained, crystalline limestone, up to 2 m in length. The structurally higher part of the formation comprises black, glassy chert in tabular beds, commonly 100 mm thick, but ranging in thickness from 50 mm to 0.3 m, with sporadic thin black shale interbeds. Some chert, particularly the tectonised chert lenses in the southern and eastern parts of the St Mellion outlier, contains up to 40 per cent radiolaria. The chert commonly weathers to form a pale grey or greenish grey rind and is strongly jointed with orthogonal joints perpendicular to bedding. The glassy chert beds are commonly strongly tectonised with irregular, disharmonic, mesoscopic folds, pervasive jointing, steep faulting, low-angle décollements and small-scale duplex structures. Adjacent to the bounding faults, the cherts are massive, intensely quartz-veined and, in places, nodular in appearance.

Within this district, vesicular lava is seen rarely as interbeds within the formation. At Burcombe Farm [4072 6665], a thin pod of vesicular lava is present within a tectonic lens of the chert, but at Newbridge [346 679], just to the north of the district, a thick unit of vesicular lava with pillows lies in contact with bedded cherts of the Newton Chert Formation (Whiteley, 1983).

### BIOSTRATIGRAPHY

The Newton Chert Formation contains an early to late Viséan fauna (Table 12). *Posidonia* has been reported by

**Table 12**   Conodont and ammonoid biozonal schemes for the lower part of the Carboniferous.

**a**   Scheme used by Whitely (1981) for age determination of the Carboniferous rocks of the St Mellion area (quoted in the text).

**b**   Equivalent biozonal scheme from Harland et al. (1989).

**b**

| Sub-system | Series | Stages | | Ammonoid zones | Conodont zones |
|---|---|---|---|---|---|
| SILESIAN | SERPUKHOVIAN | ALPORTIAN | H₂ | Homoceras | Idiognathoides noduliferus / Streptognathodus lateralis |
| | | CHOKIERIAN | H₁ | | |
| | | ARNSBERGIAN | E₂ | Eumorphoceras | Gnathodus bilineatus bollandensis / Cavusgnathus naviculus |
| | | PENDLEIAN | E₁ | Cravenoceras | Kladognathus, Gnathodus girtyi simplex |
| DINANTIAN | VISÉAN | BRIGANTIAN | V3c | Hypergoniatites | Gnathodus girtyi collinsoni |
| | | ASBIAN | V3b | | |
| | | HOLKERIAN | V2b V3a | Beyrichoceras / Goniatites | |
| | | ARUNDIAN | V1b V2a | | Gnathodus texanus |
| | | CHADIAN | V1a | Merocanites / Ammonellipsites | Scaliognathus anchoralis, Dolignathus latus |
| | TOURNAISIAN | IVORIAN | Tn 3 | | Gnathodus typicus |
| | | HASTARIAN | Tn 2c | Protocanites / Pericyclus | Siphonodella crenulata, S. Isoticha |
| | | | Tn 2b | | Siphonodella sandbergi |
| | | | Tn 2a | | Siphonodella duplicata |
| | | | Tn 1b | Gattendorfia | Siphonodella sulcata |
| | | FAMENNIAN | Tn 1a | Wocklumaria | Siphonodella praesulcata |
| | | | | | Bispathodus costatus |

**a**

| Sub-system | Series | Stages | Ammonoid zones | | Conodont zones |
|---|---|---|---|---|---|
| SILESIAN | NAMURIAN | H₂ | | | I. noduliferus / S. lateralis |
| | | H₁ | | | |
| | | E₂ | | | C. naviculus |
| | | E₁ | | | I. nodosus |
| DINANTIAN | VISÉAN | V3 | Goniatites | G. striatus / G. crenistria | G. bilineatus |
| | | V2 | | A. nasutus | |
| | | V1 | | A. kochi | G. texanus |
| | TOURNAISIAN | Tn 3 | Ammonellipsites | | S. anchoralis, D. latus / G. typicus |
| | | Tn 2c | | A. plicatilis | S. crenulata |
| | | Tn 2b | | A. princeps | |
| | | Tn 2a | | | S. sanbergi / S. duplicata / S. sulcata |
| | | Tn 1b | Gattendorfia | G. crassa / G. subinvoluta | |
| | | Tn 1a | Wocklumaria | W. sphaeroides / W. subarmata | B. costatus |
| | | Fa 2d | | | |

Whiteley (1983) from cherts at Freeres Farm [3533 6568] together with the conodont *Gnathodus* cf. *texanus* in a track between Newton Barton and Newton Mill [3455 6632] on the Newton Ferrers Estate. *Posidonia* sp. and *Neoglyphioceras spirale* were also reported from the adjacent area on Herod Down [356 646] and from Halton Quay [416 657] by Ussher (1907). In a track near Tor Farm [3598 6508], Whiteley (1983) recorded a varied assemblage of late Viséan age comprising *Neoglyphioceras spirale* and *Posidonia becheri* together with conodonts (*Gnathodus bilineatus*, *Gnathodus* sp., *Hindeodella ibergensis*, *Hindeodella* sp. and *Lonchodina* sp.) and Cyrtosymbolinid and Griffithidinid trilobites. In a small quarry at Amytree [3617 6670], cherts have yielded the conodont *Gnathodus pseudosemiglaber* (Stewart, 1981a). Whiteley (1983) found a similar fauna comprising *Gnathodus cuneiformis* and *G. pseudosemiglaber* at Viverdon Down Quarry [3739 6754] and suggested its age to be of the early Viséan *Scaliognathus anchoralis* Zone. From the same locality, Matthews (1969) reported a conodont assemblage consisting of *Doliognathus lata*, *Hindeodella segaformis*, *Pseudopolygnathus triangula pinnata* and *Scaliognathus anchoralis* together with several species of palmatolepids, interpreted by him as reworked from the Upper Devonian. Some limestone lenses in the chert at Halton Quay [4166 6573] were reported to contain the foraminifera, *Endothyra*, *Trochammina*, *Dentalium* and *Nodosinella* (Fox, 1896), and conodonts of the species *Gnathodus bilineatus*, *G. cuneiformis*, *G. girtyi*, *G. pseudosemiglaber*, *G. semiglaber*, *G. texanus* and *Paragnathodus commutatus*, indicative of the *Gnathodus texanus* and *Gnathodus bilineatus* zones (Chadian to Arundian) (Whiteley, 1983).

## DEPOSITIONAL ENVIRONMENT

This formation has yielded extensive marine faunas (see above). The radiolarian chert in the upper part of the formation probably accumulated from oozes either during a period of low sediment influx, or in an area of the basin, shielded from the sediment supply. In contrast the siliceous slate of the lower part of the formation had a different origin, probably by the diagenetic alteration of silty mudstone. The preservation of delicate trilobite fossils (Whiteley, 1983) within these argillaceous rocks suggests that conditions within the sedimentary basin were very quiet with few, if any, bottom currents.

The presence of chert of upper Tournaisian and Viséan age across much of the Variscides of Europe, that is in the Rheinisches Schiefergebirge (Engel et al., 1983), the northern Bohemian Massif (Behr et al., 1982), in the southern part of the Bohemian Massif (see Klominsky, 1994) and Cantabria (Julivert, 1971), suggests that the depositional conditions may be attributed to 'global' processes rather than to local basinal conditions. The presence of large volumes of contemporaneous basic lava, both in south-west England (in the Tintagel Volcanic Group of Freshney et al. (1972)) and throughout the European Variscides, indicates a possible source for the silica by way of submarine hot spring activity (see Krauskopf, 1967; Engel et al., 1983).

# SIX

# Permian

In the Rame Peninsula, rhyolite from the Kingsand Rhyolite Formation has been radiometrically dated as Early Permian in age. In Cawsand Bay, the rhyolite conformably overlies a breccia conglomerate, the Sandway Cellar Conglomerate Member.

## Kingsand Rhyolite Formation

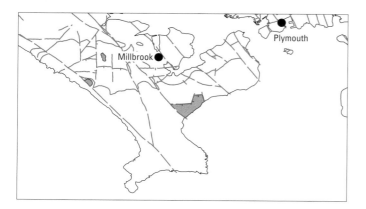

This formation comprises three discrete crops in the Rame Peninsula. The largest is near Kingsand, and smaller bodies occur at Bridgemoor [409 527] and Withnoe [404 518].

### KINGSAND

At Kingsand the formation is bounded by steep faults to the north and east, and by the Cawsand Fault, to the south-west. To the south, the coastal section provides a continuously exposed raised platform, extending from the base of the formation at Sandway Cellar beach [4410 5105] south-westwards, in the dip direction, to Kingsand Beach [4355 5055]. This principal exposure comprises mainly rhyolitic lava which overlies the Sandway Cellar Conglomerate Member.

### Sandway Cellar Conglomerate Member

This breccia conglomerate forms the reddish brown bedrock of the beach at Sandway Cellar where it is variably obscured by beach deposits. It rests with gently inclined unconformity on the steeply inclined rocks of the Staddon Formation, and basic intrusions therein. It is approximately 8 m thick and made up of thin to massive diffuse beds. Thinner beds of fine to medium conglomero-breccia, at top and bottom of the sequence, are separated by about 3 m of coarse conglomero-breccia (Plate 12). Coarse- to very coarse-grained sandstone, with dispersed clasts, forms thin discontinuous beds in the lower and upper sections, and also forms part of the matrix of the central section together with finer breccia. The rocks are largely polymict framework breccias with angular to well-rounded clasts, comprising sandstone of Staddon type, red-stained grey siltstone, hyaloclastitic basalt, limestone varieties comparable to the Plymouth Limestone Formation, quartzite, quartz vein, and rhyolite. The rhyolite clasts are similar in overall appearance to the succeeding lava and form the largest clasts, which are up to 1.5 m in diameter. In the coarsest facies, there are also blocks of the finer breccia. Some pebble imbrication is locally developed in the finer rocks.

In the finer facies, there are burrows, 50 to 70 mm in diameter and 0.5 m in length, filled with hemispherical 'meniscus' layers of aligned clasts (Professor C M Bristow, written communication, 1992). Comparable structures at Goodrington, south Devon (Laming, 1970) have been attributed to a limbed amphibian or limbless reptile by Pollard (1976) and Ridgway (1976) respectively; the former advocated the use of the trace fossil name *Beaconites* cf. *antarcticus* (Viavlov).

Locally, between the conglomero-breccia and the overlying rhyolite lava, that forms the major part of the formation, there are pockets of fine-grained, indurated, brownish red, acid tuffite with interdigitating tuff (E 67755). The tuffite comprises lithic clasts, up to 20 mm across, and dispersed epiclastic grains, similar to the finer parts of the underlying conglomero-breccia, within a very fine-grained, quartzofeldspathic, shardic dust with xenocrysts. The interfingering acid tuff is more coarsely shardic and pumiceous with corroded quartz and sericitised feldspar; the shards show the early stages of welding.

## Rhyolite

The lowest 3 m of the rhyolite at Sandway Cellar is an off-white, banded perlite (E 67753), which is hyaline, partly desilicified, limonitic and very friable. Above, it is fresh and forms a prominent upstanding feature across the foreshore. The main body of the rhyolite to its faulted upper boundary is some 100 m thick. It varies in colour between purplish pink and brownish pink, but it is altered to pale greenish grey along joints. The rhyolite locally has well-developed columnar jointing and autobrecciation. An ubiquitous, strong to diffuse banding is generally gently inclined south-westwards in conformity with the base of the flow and the underlying sediments. Flow folds, small-scale isoclines, larger scale open folds producing zones of steeper banding, and refolded folds are common. In the autobreccias, the intact rhyolite shows banding swirls. Vesicularity and siliceous nodules are sporadically developed. No evidence of internal contacts, continuous blocky breccias or weathering profiles, that might indicate that the rhyolite is a composite of more than one flow, was seen during this resurvey.

In thin section, the lava has cryptocrystalline to microcrystalline, quartzofeldspathic recrystallisation textures with sporadic feldspar microlites (for example E 60039, E 60040), phenocrysts (5 per cent) of altered euhedral feldspar (up to 3 mm long), smaller biotites and sparse euhedral and fragmented quartz. Banding is marked by variable proportions of finely divided, iron-rich, disseminated dust. Cosgrove and Elliot (1976) showed that the feldspars have features of both sanidine and orthoclase.

**Plate 12** Coarse facies of the Sandway Cellar Conglomerate Member, Kingsand Rhyolite Formation, at Sandway Cellar [4414 5106]. Looking north-westwards (GS520). Hammer is 0.30 m long.

## BRIDGEMOOR

At Bridgemoor, the rhyolite occupies a hollow at the head of a southerly draining stream, the only significant exposure being in a small quarry [4093 5257]. This aphanitic rock is pale grey with purplish red staining. It is porphyritic with feldspar (up to 10 mm long) and biotite phenocrysts, and it contains abundant clasts of country rock, up to 0.1 m long. In the quarry, flow banding is shown by vesicles flattened in the foliation and this defines an open synform with a northerly trend, parallel to the major dimension of the body. In thin section (for example E 60038, E 60041), the rock is petrographically similar to the lavas at Kingsand, apart from a more uniform microcrystalline texture with less differentiation into bands.

## WITHNOE

Rhyolite caps a hilltop at Withnoe with a semicircular crop, bounded on the south-western side by the Rame Fault. The main exposures are in Withnoe Quarry [4038 5174], which is adjacent and elongate parallel to the fault, and in a smaller quarry to the north [4036 5186]. The rock has an overall purplish brown colour, but it is finely banded in purplish red and pale yellowish brown. In the first quarry, the banding is near vertical, striking uniformly north-westwards and, in the second quarry, it is inclined moderately south-eastwards. The rhyolite is off-white to pale yellow and in places highly altered and desilicified within a zone up to 4 m wide, adjacent to the fault, that includes 1 m of brecciated rhyolite in the fault wall. Ubiquitous closely spaced jointing is gently inclined variably between west and south-west; steep widely spaced joints of differing trends are also common.

The rhyolite is porphyritic: feldspar phenocrysts, up to 20 mm across, are apparent in hand specimen. In thin section (for example E 60032, E 60034), this rock is very similar to the lava at Kingsand; it is porphyritic, cryptocrystalline and micro-crystalline. However, the banding is more clearly defined: fine laminae (commonly less than 0.05 mm wide) of quartzofelds-pathic minerals, with some micaceous alteration, alternate with laminae, rich in finely disseminated, undifferentiated chlorite and iron-rich compounds. Locally, these laminae have a glassy texture. The banding shows complex eddy structures in the shadows of phenocrysts of altered euhedral feldspar and biotite.

In Withnoe Quarry, the rhyolite is intruded by quartz-feldspar porphyry dykes, elvans (Figure 23). The constituent rock is pale pinkish grey to yellowish grey and has a coarser texture and higher proportion of quartz, feldspar and biotite phenocrysts than the rhyolite. The dykes, up to 1.5 m thick, anastomose and interconnect through the rhyolite; the principal trends are parallel and perpendicular to the banding, and the Rame Fault. The dykes do not have significantly finer grained chilled margins against the rhyolite, with which they share the gently inclined jointing. In thin section (for example E 60033, E 60036) the rock is seen to comprise microcrystalline quartz and feldspar with finely disseminated ore minerals and about 20 per cent phenocrysts. Quartz phenocrysts (up to 2.5 mm long) that are euhedral, corroded embayed or rounded euhedral, make up half of the phenocryst content. The remainder is euhedral orthoclase as phenocrysts up to 10 mm long.

### GEOCHEMISTRY AND GEOCHRONOLOGY

The major element, trace element and Rare Earth Element geochemical analyses of the rhyolites at the three localities, and of the elvans at Withnoe, show similar results (Figure 24a, b, c), indicating their close genetic and emplacement affinity.

Radiometric dating by Rundle (1981) of the lavas at Kingsand, using the K–Ar method on biotites, yielded a mean age of 289 ± 4 Ma. This result places these rocks in time just above the Permo-Carboniferous boundary (Harland et al., 1989; Roberts et al., 1995) in the Early Permian.

Darbyshire and Shepherd (1994) carried out an Sm–Nd isotope investigation of the granites and elvans of south-west England. The $\Sigma_{Nd}$ (T) for the main plutons range from -4.7 to -7.1 indicating significant differences between individual granites. Initial Nd isotope signatures of the Kingsand/Withnoe rhyolites and elvans (-6.5 to -7.2) suggest a close relationship with the Bodmin granite (-7 at T = 290 Ma) rather than with the

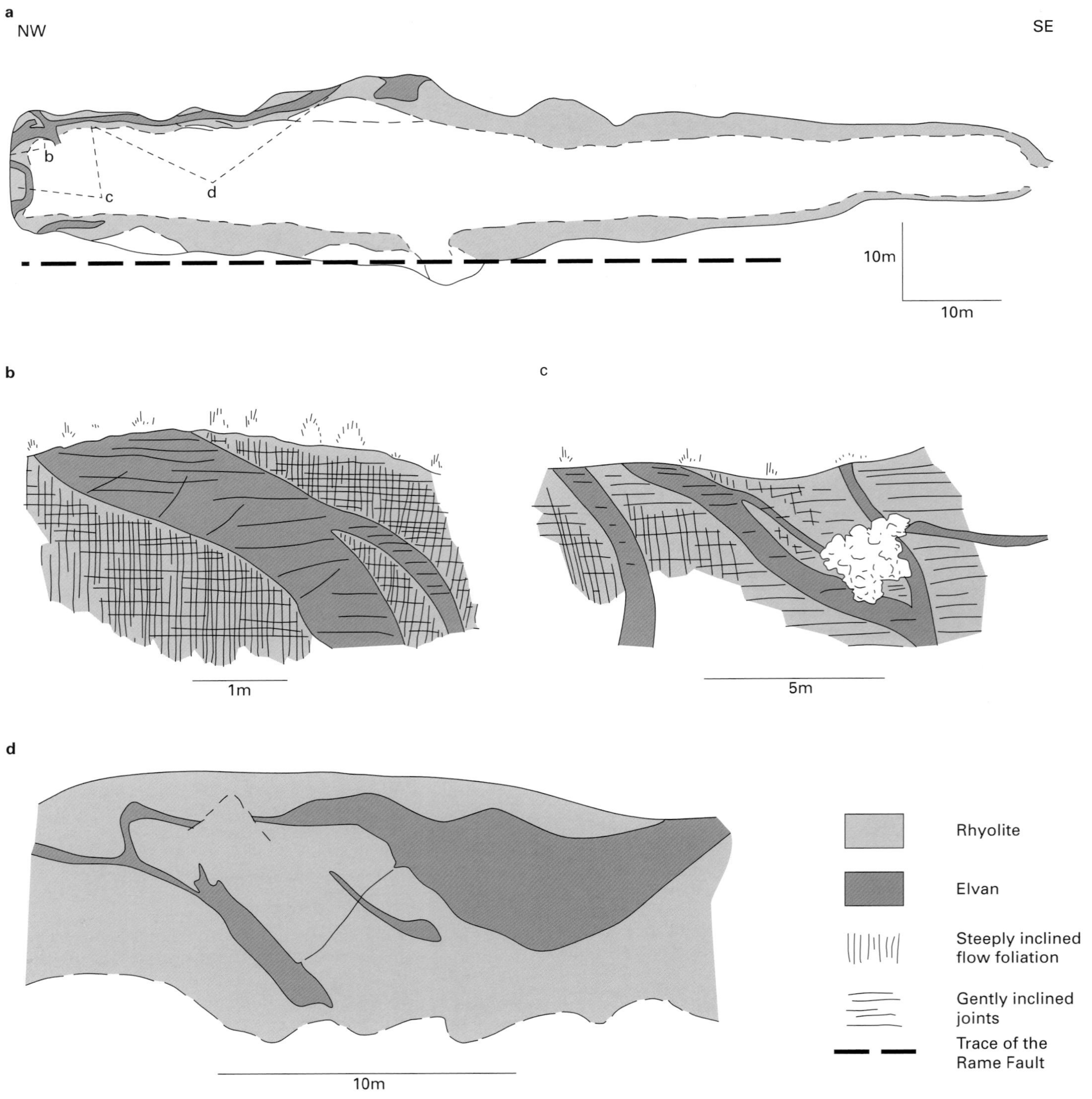

**Figure 23** Field sketches of Withnoe Quarry showing rhyolite and elvans.

a. Plan view of the quarry showing location of the elvan dykes at the western end. Locations of b, c and d also shown
b. Bifurcating elvan within rhyolite in northern corner of the quarry. Vertical fabric is flow foliation, subhorizontal traces are joints
c. Interconnecting dykes in northern corner of the quarry
d. North-eastern face of quarry, subparallel to main trend of a dyke and steep flow foliation in rhyolite, showing elvan apophyses towards the viewer from the side of the dyke

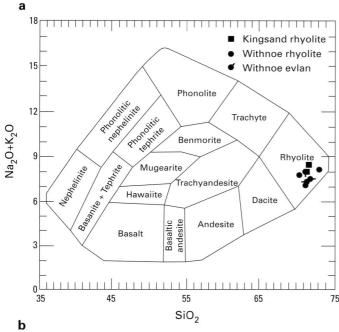

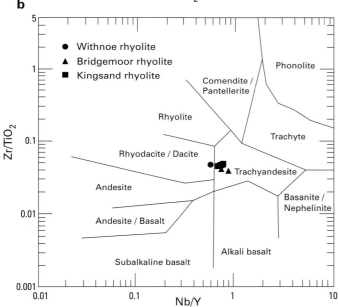

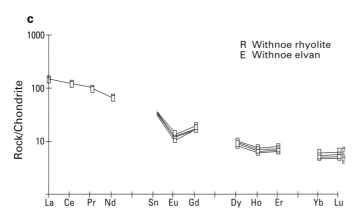

Dartmoor pluton which has a less negative $\Sigma_{Nd}$ value (-4.7 at T = 280 Ma).

EMPLACEMENT AND SETTING

The Sandway Cellar Conglomerate Member is a framework breccia conglomerate, typical of rapid deposition close to source. The polymict nature of the deposits points to fluvial emplacement rather than proximal fan deposition. The included large blocks of rhyolite, similar to the overlying lava, may relate to the erosion of an earlier lava of the same episode, possibly facilitated by penecontemporaneous faulting.

At Kingsand, the dip of the lava into the Cawsand Fault, and the largely fault-bounded nature of the body, suggests that the considerable thickness of the lava may result from ponding in a fault-guided sag-basin during regional extension. The autobrecciation at various levels indicates possible carapace formation during emplacement. There is no direct evidence of the nature of the rhyolite at Bridgemoor, but the location and gently inclined banding suggest that it is an outlier of lava, rather than intrusive. The rhyolite at Withnoe was considered to be a plug by Cosgrove and Elliott (1976). The vertical banding, adjacent and parallel to the Rame Fault, the inclined banding away from the fault, and the sub-horizontal basal contact elsewhere, point to the fault being the conduit for a tholoidal (dome-shaped) igneous body.

Worth (1891) and Hobson (1892) both observed an association of different rock types at Withnoe Quarry, the latter recording the presence of four dykes. Tidmarsh (1932) noted that the rhyolite was intruded by elvan dykes, but more recent investigators overlooked this. The geometry of the elvans and the anastomosing and interfingering character of the interconnecting dykes indicate high-level intrusion in a strain regime when the parallel Rame Fault was in extension. The similar chemistry of the elvan dykes and rhyolite suggests that, although slightly later than the rhyolite, the dykes were feeders for similar extrusives.

Scrivener et al. (1995) have linked various expressions of Permo-Carboniferous acid extrusive rocks in southwest England to different major granite cupolas. In the case of the rocks in the Rame Peninsula, there is close genetic relationship with the Bodmin Moor Granite, based on their chemistry and age. The Rame and Cawsand faults both pass north-westwards to that granite and it seems probable that the Rame/Portwrinkle fault system acted as a conduit for magma, linking with the cupola of the Bodmin Moor Granite.

**Figure 24**  Geochemical diagrams showing data for rhyolites and associated elvans of the Kingsand Rhyolite Formation.

a.  Total alkalis versus silica (TAS) diagram modified from Cox et al. (1979)
b.  Discriminant diagram of Winchester and Floyd (1977)
c.  Chondrite normalised Rare Earth Element diagram

SEVEN

# Cainozoic

No deposits within the Plymouth district have been dated as being of Palaeogene or Neogene age. The resurvey has revealed, however, the metre-scale sculpturing of the topography during sea-level retreat, as elsewhere in south-west England (cf. Leveridge et al., 1990), which gave rise to a succession of wave-cut platforms, and to the formation of sea caves within the Devonian limestone, during these periods.

## QUATERNARY

Quaternary deposits are widely scattered through the district and these comprise fluvial, aeolian, estuarine and marine sediments. Solifluction and residual deposits, termed head, also occur along the coast, in river valleys and on hillslopes. The marine sediments include raised beach gravels, localised shoreface and beach sands, and the deposits of tidal river, creek and saltmarsh. Aeolian deposits are represented by small areas of blown sand near the coast, as in the valley south of Tregantle Fort [386 533]. The fluvial deposits consist solely of alluvium, predominantly gravel, silt and clay. Submarine forests in the present-day intertidal zone have been described previously in the district but were not seen during this resurvey. Cave deposits of Quaternary age consisting of stratified deposits bearing mammalian remains are present in caves within the Plymouth Limestone Formation. Landslips are present at various locations along the coast.

Considerable climatic variation occurred in the British Isles during the Quaternary when glacial and periglacial phases were separated by periods of temperate (interglacial) climate. Although during the Anglian, Wolstonian and Devensian glacial stages the British Isles were covered extensively by ice, this district lay well beyond the southernmost limits of these ice sheets. These climatic changes gave rise to variable sea levels, in response to the changing volume of water locked up as ice, and these formed raised beach features and buried river channels along the coasts of southern Britain. In the Falmouth district of west Cornwall, a series of such beach features at about 35 m, 20 to 25 m, 10 to 15 m and 2.5 to 5 m above OD have been identified as being of probable Pleistocene age (Leveridge et al., 1990). The presence of unconsolidated sand and clay of Oligocene age at St Agnes (Atkinson et al., 1975) and of Pliocene age at St Erth (Leveridge et al., 1990) indicates that elevated Pleistocene sea-levels could not have been higher than 40 m above the current OD. This limit is supported by recent research into Pleistocene raised beaches between Sussex and Pembrokeshire which has revealed five separate raised beaches below 40 m, the lowest four of which were correlated all along this stretch of coast (Bowen, 1994).

Buried river channels associated with periods of low sea level were recorded by Codrington (1898) beneath the Tamar and Plym estuaries (Table 13). More recently, seismic profiling of the Hamoaze and The Sound (Eddies and Reynolds, 1988) revealed the presence of a buried valley between 40 and 44 m below OD, in which four levels of erosion were identified. The last phase of downcutting is thought to have occurred in Late Devensian times. As sea level rose during the subsequent Holocene climatic amelioration, sedimentation occurred in the buried valleys. In the recent past, mine tailings were fluvially redistributed in many of the valleys west of the Lynher River. This process continues at a very slow rate at the present-day.

### Caves and cave deposits

No dateable cave deposits have been recorded in this district but a number of deposits related to both glacial and interglacial periods have been identified regionally (see MacFadyen, 1970; Oldham et al., 1978). These deposits provide evidence of the variations in the Pleistocene climate but little evidence about sea levels because they are all subaerial deposits, postdating the time of cave formation. The caves in which they are found, however, display evidence of a complex history of formation in both subaqueous and subaerial environments and hence provide some clues to the history of sea level variation in the district.

The caves within the Devonian limestone of Devon occur at a variety of heights relative to the present OD, from 70 m below sea level to 190 m above it (Oldham et al., 1978). The caves comprise two distinct elements: horizontal caverns and steeply inclined to near-vertical rifts. Many of the caves are made up of one or the other of these elements, but cave systems in which both cave types occur are not uncommon. The distribution of the number of horizontal caves at any particular height is not uniform (Figure 25). The main features of the distribution pattern are, however, masked by a general bias towards levels corresponding with the 50 foot and multiples of 100 foot contour intervals. When the data are reappraised to approximate all of the cave heights to the nearest 50 foot level, the distribution shows three discrete maxima at around 15 m ± 8 m, 60 m ± 8 m and 135 m ± 8 m.

The cave systems in the Plymouth Limestone Formation, however, are limited to a maximum level of about 50 m above OD by the topographical expression of the limestone crop. Within this group, there is a distinct

**Table 13** Depths of buried river channels in the Tamar and Plym estuaries (from Codrington, 1898).

| Locality | Level | Notes |
|---|---|---|
| Laira Railway bridge | - 26.5 m† | 64.62 m in centre of river not probed |
| Cattewater | - 19.5 m‡ | |
| Sutton Pool | - 18.3 m† | |
| Millbay docks | - 22.2 m† | 16.76 m of silt present at outer end of docks |
| Millbay to Drake's Island | - 32.9 m† | |
| Opposite Eastern King Point | - 42.1 m | |
| Devil's Point to Wilderness Point | - 36.6 m | |
| Weston Mill Creek railway bridge | - 20.1 m‡<br>- 18.3 m† | |
| Hamoaze (near Weston Mill Creek) | - 21.9 m‡ | 21.96 m silt |
| Tavy railway bridge | - 20.7 m‡ | 20.73 m of silt underlain by stiff yellow clay with small 'granite' boulders |
| Saltash bridge | - 23.2 m† | 7.32 m of silt |
| Forder Lake | - 20.1 m† | 21.34 m of silt |
| Wiveliscombe Lake | - 14.0 m‡ | 17.07 m of silt |
| Notter Creek railway bridge | - 14.0m† | 16.76 m of silt |
| St Germans River | - 12.5 m† | 15.24 m of silt |

† low water spring tides          ‡ below low water

†‡   Reference points are those of Codrington (1898). Spring tides are those with maximum amplitude and thus lower than normal.

tendency for cave formation to have taken place around the 25 m level.

Where rifts occur within cave systems, they show a large range of vertical extent (Figure 25) and appear to interconnect all of the cave levels. Moreover these rift caves form the deepest known cave systems in south-west England. The submarine caves in the Plymouth Limestone Formation, extending down to 70 m below OD all appear to be, in part at least, rift systems (BGS archive data; Oldham et al., 1978).

### Age of formation

In west Cornwall, the maximum height of sea level during the Pleistocene was approximately 35 m above OD (Leveridge et al., 1990). The presence of undisturbed, loose sand and clay of late Pliocene age (Mitchell, 1973) at St Erth, at 45 m above OD (Leveridge et al., 1990) suggests that most of the Devon caves that lie above 40 m are likely to be of Neogene age or older. Although this conflicts with the general assumption that caves within south-west England are entirely Pleistocene features (Cullingford, 1982), open caves of Neogene age are known in other parts of the world (see Osborne, 1983, 1986). The maxima in the height distribution pattern of the Devon caves therefore appear to correlate with the more pronounced sea level still-stands within the Palaeogene and Neogene which gave rise to the regional erosion platforms at 130 m above OD and 70 m above OD (see Leveridge et al., 1990). Caves formed at even higher levels in the Kingsteignton area, at approximately 190 m above OD (Oldham et al., 1978), are partly filled by unconsolidated sand similar to that of the unconformably overlying Upper Greensand and therefore may be of Cretaceous age.

### Formation mechanism

Horizontal caves of presumed Palaeogene and Neogene age in Devon generally exhibit evidence of both phreatic and vadose mechanisms of formation, modified in places by substantial breakdown of the cavern roof (see MacFadyen, 1970; Oldham et al., 1978; Jean, 1984). In these caves, the phreatic features are the earliest and appear to relate to their formation by flowing water, possibly tidally influenced, below the water table (see Bogli, 1980). Vadose features such as vadose notches in cave floors, flowstone floors and stalagmites and stalactites postdate the phreatic features. Breakdown of the cavern roofs occurred subsequent to the onset of vadose conditions (Plate 13). Although localised freshwater lakes within the caves may have modified the stalactites and stalagmites by depositing flanges of calcite at the lake levels, such as 'The Little Man' in Reed's Cave (Oldham et al., 1978), there is no evidence of any subsequent renewed phreatic activity. The implication of this sequence of events is that, from the end of the Cretaceous until the start of the Pleistocene, sea levels in south-west England fell progressively.

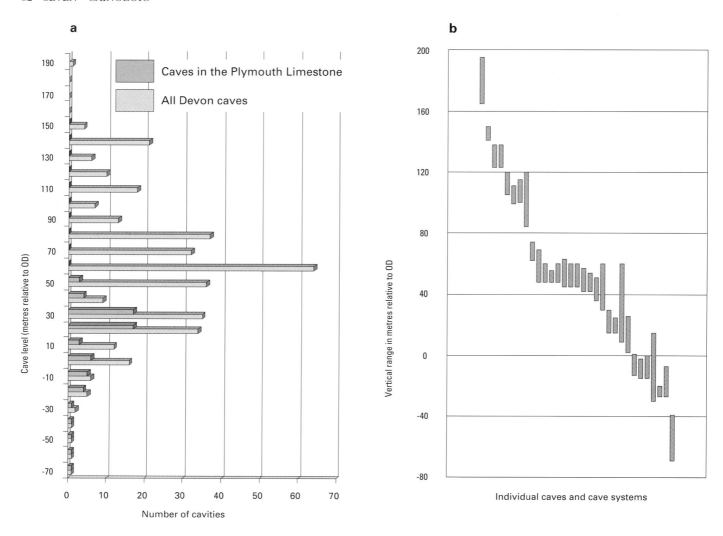

**Figure 25**   Plymouth caves in relation to limestone caves of Devon.

a.   The levels of cave systems in the Devonian limestone in south-west England

b.   Vertical ranges and heights of rift cave systems in Devon (data from Oldham et al., 1978; BGS archives)

The onshore caves within the Plymouth Limestone Formation all lie below 35 m above OD and therefore fall within the range of regional Pleistocene sea levels, as established above. Because the Palaeogene and Neogene caves record a progressively falling sea level, all of the caves lying below the lowest Neogene sea level are of Quaternary origin and therefore cannot be exhumed older cave systems. All of the caves presently above sea level exhibit signs of phreatic formation (Jean, 1984) and were probably formed during Pleistocene interglacials when sea levels were higher than at present. However, the exposed caves and rifts also display features such as flowstone floors, stalactites and stalagmites (Jean, 1984) indicating a later period of vadose activity. The deposits within these interglacial caves contain remains of cave bear, grizzly bear, bison, red deer, horse and slender-nosed rhinoceros at Oreston (Worth, 1931) and hyena, wolf, rhinoceros, lion, boar, man and reindeer at

Cattedown Cave (Lattimer, 1961) to the east of the district. The caves at Stonehouse (now lost) were reported to have yielded remains of rhinoceros, red deer, ox, horse and ass (Geach, 1936).

Caves below present sea level in the district are, for obvious reasons, not well known. Evidence from divers (Oldham et al., 1978) and from borehole records (BGS archive data) suggests that these caves are in part rift systems. These rifts display evidence of vadose activity in flowstone coatings, but may have formed initially in a phreatic environment as interconnections between horizontal systems as the sea level fell during glacial periods. Hollows in the beds of the River Tamar (Eddies and Reynolds, 1988) and the River Plym at Cattedown (BGS archive data) at approximately 40 m below OD may be collapsed caverns or swallow holes formed when sea level was 40 m lower than at present. The entrance to the Millbay blue hole is reported by

**Plate 13** Calcite sandstone infilling of a cave [4605 5332] near Western King Point in a limestone sequence that is gently to moderately inclined south-eastwards (right). The horizontally laminated sandstone rests on, and is overlain by, collapse breccias (GS521).

Oldham et al. (1978) to be 30 m below OD and this may form part of a more extensive horizontal system at this level. The presence of a shaft in the Millbay blue hole indicates phreatic or vadose solution of rifts during the Pleistocene when sea level dropped as low as 70 m below OD.

Many of the Pleistocene caves at lower levels have undergone extensive modification as sea level varied between the glacial and interglacial periods. The submarine caves in Millbay are currently undergoing phreatic alteration due to tidal flow through the caves (see Oldham et al., 1978), and the caves at sea level around The Hoe have probably been extensively altered by wave action (see Oldham et al., 1978; Jean, 1984).

**Marine deposits**

RAISED BEACH DEPOSITS

Residual patches of raised beach deposits commonly occur on the coast of this district, but are too small to be shown on the map face. The heights of the various beach levels correspond to those noted in the Falmouth district (Leveridge et al., 1990) at 2.5 to 5 m, 10 to 15 m, 20 to 25 m and at about 35 m. The first described deposit in this district (Hennah, 1816) was of sand and waterworn pebbles at The Hoe, at a height of 4.5 to 5.5 m above high water. This may equate in level with that which is currently exposed at 2.5 m above OD in the cliff beneath the walls of The Citadel [4794 5370] (Plate 14). Here there are cobbles and pebbles of locally

**Plate 14** Raised Beach on the 2.5 to 5 m platform near the Royal Citadel [4794 5370]. Looking northwards (GS522). Hammer is 0.30 m long.

derived limestone, some Lower Devonian sandstone, siltstone, tuff and lava and Permian rhyolite, in a matrix of calcareously cemented sand. Some broken shell debris is also present and the deposit fills cavities along faults which form a north-westerly trending fracture zone through the bedrock. Other examples of the 2.5 to 5 m raised beach occur between Picklecombe Fort [456 515] and Kingsand [434 505], Crowstone Cliff [3910 5228], west of Seaton [2990 5425] and east of Looe [261 534].

At The Hoe [475 538], a second raised beach platform is present at a height of 10 to 15 m although road construction has obscured any beach deposits on it. A higher platform is present at 20 to 25 m above OD in the cliffs between Rame Head [408 481] and Penlee Point [443 487]. A mantle of head is draped over it and no beach deposits were observed during this resurvey. At an even higher level of about 36.5 m, Whitley (1882) described a deposit of boulders and pebbles in a matrix of clay with white and red sand, exposed during hotel construction on The Hoe. Fossil shells extracted from the clay matrix clearly indicated a marine origin. According to Worth (1882, 1888), 40 per cent of the clasts in this deposit were of flint and other pebbles comprised schorl and granite.

These raised beaches are all thought to have formed in interglacial stages. In this district, deposits at the 2.5 to 5 m level, near Polhawn Cove [4211 4945], comprise 3 m of pebbly sand (beach deposits) overlain by 4 m of blown sand and 10 m of solifluction material (head), indicating that the beach was abandoned prior to the onset of cold conditions. Deposits at the same level in the Falmouth district were correlated with the Ipswichian interglacial by James (1981). Raised beach deposits lying on platforms between 4 and 7.5 m and at 25 m above OD in the Isles of Scilly were assigned by Mitchell and Orme (1967) to two separate stages, the Ipswichian and Hoxnian interglacials. Hoxnian ages for raised beach deposits in south-west England have also been suggested by Mitchell (1960) and Stephens (1966) and Ipswichian ages by Zeuner (1959), Bowen (1969, 1973), Kidson (1971) and Kidson and Wood (1974) although the height of the deposits dated is not clear. The preservation of higher raised beach deposits may reflect a generally falling sea level for successive interglacials and suggests that the deposits on the three raised beach levels above 5 m formed during at least three earlier interglacial periods of the Pleistocene.

SHOREFACE AND BEACH DEPOSITS

These deposits are only sporadically distributed along the local coastline. Small patches of sandy beach material occur near the mouths of the rivers Looe and Seaton, whereas an almost uninterrupted sandy beach, known in part as the Long Sands, forms the foreshore between Portwrinkle and Rame.

TIDAL RIVER AND CREEK DEPOSITS

The main channels of the rivers Lynher, Tamar and Tavy estuaries are bordered by considerable areas of mudflats which are covered at high tide. These areas are formed of silt and silty mud, commonly having a considerable organic content. Some of these deposits, especially those that lie close to bedrock, include a significant amount of locally derived rock debris. Tidal winnowing of these deposits has in places led to a concentration of the rock debris, shells and modern waste in localised shoals and beaches. There has been a long history of using the local estuaries as a dumping ground for all forms of waste, so that for example at Sand Acre Bay [4190 5747], the mud is heavily charged with broken glass and ceramic debris, much of it Victorian in age.

SALTMARSH DEPOSITS

The main areas of saltmarsh occur on the fringes of the estuaries of the rivers Lynher, Tamar and Tavy. The most extensive saltmarshes are present on the River Tamar west of Bere Ferrers [440 635] and to the north of Clifton [425 650]. These areas of brown and grey clayey silt are inundated at high water by all except the lowest of neep tides.

SUBMERGED FOREST

Although no occurrences of submerged forest were observed during this resurvey, De la Beche (1839, p 419) recorded 'a small portion' of submarine forest at Millendreath, near Looe. A 'submarine forest' is noted at the mouth of the Polperro River on the 1888 edition of the 1:10 560 Ordnance Survey Cornwall County Series Sheet LII. S.E. but apparently this was not observed by Ussher in the 1899 survey. These occurrences probably lie in the intertidal zone and may equate with an occurrence of submarine forest in the Penzance district, at Long Rock, Mount's Bay, where it has been radiocarbon dated as 4278+/-50 years BP (Goode and Taylor, 1988). Forests preserved in the present intertidal situation provide evidence of a temporary pause in the overall rise of Holocene sea level.

OFFSHORE DEPOSITS

The offshore deposits of the district are displayed on the 1:250 000 scale map of sea bed sediments and Quaternary geology (BGS, 1987). These deposits consist mainly of sandy gravel and sand between Downderry and Rame Head. Muddy sand occurs in Cawsand and Jennycliff bays and a deposit of gravelly muddy sand lies south of The Hoe, between Barn Pool and Mount Batten Point.

## Fluvial deposits

ALLUVIUM

Most of the major river systems of the district are tidal and are floored by tidal river or creek deposits, and so river alluvium occurs only in tributary streams. Some of the larger river valleys, such as that of the Lynher River, are flat-floored, rather than 'V-shaped', indicating some form of wave cutting in their formation. In such valleys, the alluvial deposits of silty clay and thin gravel beds, despite their broad spread across the valley floor, are generally thin, and the river bed is on bedrock.

Elsewhere, the alluvium comprises a thin cover of reworked head deposits which may pass laterally into the head deposits fringing the valley slopes.

## Aeolian deposits

### BLOWN SAND

A small area of blown sand occupies the valley to the south of Tregantle Fort [386 533]. This deposit is of recent origin. Where currently being redeposited by storms, the sediment is an off-white to pale yellow, fine- to coarse-grained quartz sand. In areas where it is not subject to reworking, it is pink and reddish brown.

## Mass movement and residual deposits

### HEAD

The term 'head' was first used by De la Beche (1839) to describe deposits of angular fragments resting on raised beaches in south-west England. The term is currently used to describe poorly or non-stratified deposits of clay, silt, sand and locally derived angular lithic clasts, which were transported downslope by solifluction under periglacial conditions. Much of this head, which is up to 30 m thick, was derived from pre-existing weathering products that formed a regolith, up to 2 m thick. There is a gradational passage between the two materials. Solifluction took place under repetitive freeze–thaw climatic conditions which enabled unconsolidated soil and rock debris to move down slopes of just a few degrees, to accumulate in topographic lows. The dating of these deposits remains far from certain but, because the head overlies the lowest raised beach deposits, many authors have attributed a Devensian age to the major head deposits of south-west England. Head commonly accumulates in basins at the heads of valleys and good examples occur west of Dobwalls [206 647], north of Liskeard [249 665] and west of Trerulefoot [318 588]. Probably the best exposures of head are to be seen along the coast around Talland Bay [225 515], west of Hannafore Point [250 520], at Portwrinkle [358 538], between Rame Head [418 481] and Penlee Point [443 487], and between Kingsand [434 505] and Picklecombe Fort [456 515].

### LANDSLIP

Landslips are not a common feature of the coastline of this district and they are rare inland. The largest slip occurs between Millendreath [269 539] and Seaton [304 544] where bedding and cleavage are subparallel in the Dartmouth Group and dip seawards (see Chapter 2 for details).

## Artificial deposits

These are discussed in chapter 2 under the headings *Derelict ground* and *Made ground/fill.*

# EIGHT

# Igneous rocks

Igneous rocks of the district are part of province-wide expressions representing igneous activity that persisted throughout the Devonian and into the Early Carboniferous and, after a period of quiescence, reactivated in the Late Carboniferous/Early Permian. The Devonian to Dinantian igneous rocks of the Plymouth district are bimodal, but preponderantly basaltic extrusives and hypabyssal intrusions. The Lower Permian rocks are predominantly acidic, and, in contrast to elsewhere in the province, intrusions are subordinate to extrusive rocks. The composition and modes of occurrence of the igneous rocks are an integral part of the actively evolving tectonic regime (see chapter 11). The basaltic rocks are present in all the sedimentary basins, albeit in minor proportions in the Looe Basin and in the Tavy Basin. The major expression of igneous activity in the district is within the Middle and Upper Devonian rocks of the South Devon Basin. There is a close association with the Plymouth High where volcanicity persisted in its vicinity from the Eifelian through the Givetian and into the Frasnian. In the St Mellion outlier, there are minor occurrences of the widespread Lower Carboniferous volcanic rocks of the southern Culm Basin.

## IGNEOUS ROCKS IN THE LOWER DEVONIAN

### Extrusive rocks

Volcanic rocks constitute a significant part of the Bin Down Formation but in poorly exposed and featured ground, in which surface brash indicates an intimate association of volcanic and sedimentary rocks, expressions of vesicular basic lava, tuff and tuffite have no great mappable continuity. Fine lithic crystal tuff, within an essentially silty mudstone sequence, is traceable over a few hundred metres near Wilton [310 584] towards the eastern end of the formation crop, and nearby in the valley of the River Seaton [305 581] bedded tuff passes into mudstone tuffite, over a distance of 200 m from west to east. To the west in a disused quarry [2522 5750] near Tregarland Tor a thin bed of silicified, coarse-grained, lithic, crystal tuff is isolated within a thick sequence of medium to thick beds of quartzose sandstone. Between these areas, a similar association is present in a disused quarry [2757 5758] near the clubhouse of the Looe Golf Club which is situated on a prominent east-north-east-trending ridge that is underlain by lava, tuff and sandstone. The Widegates Borehole [2748 5747] sunk nearby established a basaltic volcanic sequence, some 40 m thick, in the upper part of the Bin Down Formation. That sequence comprises thin to thick, diffusely bedded tuff, with sporadic interbedded tuffite, passing into massive hyaloclastite and coarse brecciated

basalt towards the top. The tuff varies from basaltic clast supported and ungraded to matrix supported and graded, and exhibits mass flow and turbidity flow characteristics. Available evidence points to pre-burial silicification of the volcanic pile, with cryptocrystalline quartz replacing host-rock groundmass and microcrystalline infill of amygdales. In the top few metres of the sequence only ghost textures (e.g. E 70446) and trace element/Rare Earth Element geochemistry indicate the original basaltic composition. In the previous survey these rocks were classified as felsite on the map.

Elsewhere in the Lower Devonian, extrusive volcanic rocks are very sparsely developed with a few minor occurrences of basaltic lava [e.g. 2320 5475] in the marine Bovisand Formation and volcaniclastic rocks in the Whitsand Bay Formation. Notable amongst the latter are laminae and thin diffuse beds of mudstone tuffite interbedded with green silty mudstone near Captain Blake's Point [4184 5006]. The volcanic components of the tuffite include feldspar, quartz crystal fragments and accretionary lapilli, suggestive of emergent acidic volcanicity.

### Intrusive rocks

Intrusive rocks within Lower Devonian sequences are of Early Devonian age and younger. Dykes intruding the Whitsand Bay Formation in the vicinity of Polperro and Looe are interpreted to be penecontemporaneous with the sedimentary rocks. Just east of Polperro harbour [2134 5080] a dolerite dyke intrudes a sedimentary growth fault (Figure 15) and the contact of the dyke with the surrounding green slaty mudstone is frilled. On the foreshore [2575 5275] near Hannafore, a 2 m-wide dyke of pale greenish cream coloured dolerite displays similarly highly irregular, interfingering contacts with host grey slaty mudstone and sandstone.

Minor intrusions up to a few metres in width elsewhere have either sill or dyke form. They are recorded where there is exposure, along the coast, in the deeply cut valleys, and in a few small quarries [e.g. Ladyswell Quarry 3739 5485]. Distribution is sporadic along much of the coast but at Rame Head, from Sandway Cellar [441 511] to Hooe Lake Point [449 513] and between Picklecombe Point [456 515] and Redding Point [460 518] there are intrusion swarms. Between Sandway Cellar and Hooe Lake Point some of the dykes are multiple, with internal chilled margins, and show a marked contrast in amygdale size at the contacts. The dykes are up to 5 m in width. Commonly they trend parallel to either strike or dip of the country rocks, as near Redding Point where dykes are predominantly perpendicular to $D_1$ fold axes. Contiguity of $S_1$ in dykes and country rocks there suggests intrusion into a–c extensional fractures late in

$D_1$ (see chapter 10). Similarly cleaved dykes near Rame Head associated with $D_1$ strike-slip faulting indicate intrusion late in that deformation.

The Longstone [3375 5363] is a small promontory formed by a body of gabbro in excess of 50 m thick and extending about 400 m along strike (Barton et al., 1993). It is conformable in part with a sheared lower margin dipping southwards parallel to cleavage, but an exposed side margin is irregular and oblique to bedding in the country rocks. The upper margin is not exposed. A differentiated mafic highly transgressive sill or dyke, some 15 m thick, interconnects with the base of the sill, suggesting a laccolithic form to the body. The main body comprises clinopyroxene crystals (up to 8 mm) in ophitic relationship with sericitised feldspars (E 66126–7); the igneous texture is disrupted by crystal fracturing and anastomosting chlorite-filled shear planes in the few metres adjacent to the basal contact.

## IGNEOUS ROCKS IN THE MIDDLE AND UPPER DEVONIAN

### Extrusive rocks

The two major volcanic expressions in the district were apparently largely separate geographically but overlapped temporaly. To the south, volcanic rocks closely associated with the Plymouth Limestone are essentially of Middle Devonian age but extend into the Late Devonian. To the north, in the belt extending westwards from the Compton area of Plymouth through Saltash and Landrake to the outskirts of Liskeard, the volcanic rocks extend up from the Givetian through the Late Devonian.

The southern expression is in large part closely linked with the Plymouth High. Bedded tuff and hyaloclastite of Eifelian age are present in the upper part of the Jennycliff division of the Saltash Formation and in the Faraday Road Member of the Plymouth Limestone Formation to the north. Gravity and magnetic profiles of the Plymouth Limestone modelled by Shelton (1987) suggested that up to 300 m of volcanic rock lay beneath the limestone. Within the limestone formation itself, late Eifelian and Givetian tuffs and lavas are present on Drake's Island and late Givetian and Frasnian tuffs, hyaloclastites and lavas are present at Mount Wise and Mount Batten. Volcanicity over a considerable period produced deposits that were an integral part of the developing edifice of the Plymouth High, and were possibly significant in forming elevated areas on which bioherms developed. The St Germans Tuff Member, of Givetian age, equivalent in age to a thick massive and bedded hyaloclastite sequence on Drake's Island, was the only significantly major expression to extend northwards out into the South Devon Basin from the Plymouth High.

On Drake's Island [469 528] the structurally uppermost tuff sequence, considered to be part of the Jennycliff division, is thinly to thickly bedded, with lapilli showing normal grading, and dispersed blocks up to 0.3 m. Carbonate veining with metallic sulphides, and hematite concretions are common. The tuff consists chiefly of cleaved comminuted hyaloclastite and lapilli of basic pumice and lava with calcite amygdales (E 66428–9).

The St Germans Tuff Member is variable laterally. It comprises largely discontinuous masses of vesicular basaltic lava and hyaloclastite in the Insworke area, passing to massive and bedded hyaloclastite with overlying interbedded lithic tuff, chert and grey mudstone in the St John area, and up to 175 m of lava and tuff in the St Germans area. The cleaved basic tuff at St Germans [e.g. 3639 5736] comprises fragments of basaltic lava and hyaloclastic debris in a carbonate-cemented, chloritic matrix (E 66402). At the same locality, basic lava consists of largely secondary albite, chlorite, sphene and carbonate replacing primary minerals (E 66403).

Basaltic lava and associated dolerite are worked at Lean Quarry, near Horningtops [271 608]. The stratigraphical position of the lava is unknown but, by restoration of displacement on the Portwrinkle Fault and correlation with limestone in the Polbathic area, a Givetian age can be suggested. The lava, which is well cleaved, commonly comprises well-developed small pillows up to 0.5 m, with chlorite and calcite infilling vesicles (E 67671–2). The intimate association of the dolerite, which consists of granular and subophitic plagioclase, augite and opaque iron oxide [E 25523, E 33790], points to the possibility that it represents magmatic feeders.

The northern belt of volcanic rocks within the South Devon Basin consists of predominantly basaltic lava, hyaloclastite and tuff, and is closely associated with intrusive bodies. It extends from Plymouth, through Saltash and westwards to Liskeard. These rocks occur largely within the Saltash Formation, but also within the Torpoint Formation in the Landrake area. They range in age from Givetian to Famennian; the main occurrence is Frasnian. The principle mode is vesicular lava, locally pillowed, as in the Wearde [429 580] coast and railway cutting sections, but massive and bedded hyaloclastite is in places predominant, as near Sawdey's Rock [364 609], and bedded basic tuff forms localised accumulations [e.g. Drake's Hill 461 583]. Although most occurrences of volcanic rocks have no great mappable continuity, there are a few areas where flows and beds at different levels amalgamate to form more extensive crops of volcanic rocks, apparently representing volcanic centres. Such centres are evident in the Compton suburb of Plymouth, about Sawdey's Rock, and near Wisewandra [348 622]. Partial sections through the Compton centre along the Manamead Road show an accumulation of diffusely bedded volcanic breccia and massive pillow hyaloclastite up to about 300 m thick. Laterally, this sequence passes into vesicular lavas and tuffs interdigitating with the slaty mudstone of the Saltash Formation. There are minor occurrences [for example at 4391 5891] of acidic vitric tuff (E 59988, E 59989) observed in association with these basic rocks.

The valley of the River Lynher provides a dip section through numerous basaltic lavas and clastic volcanic rocks of the northern group between Poldrissick [382 591] and the Notter area [385 612]. The lavas are predominantly massive, but some are pillowed, with pillows locally up to a few metres across. Typically in thin section (e.g. E 66424) they comprise a flow-aligned groundmass of plagioclase laths (now albite), sphene, interstitial chlorite with carbonate, and sparse phenocrysts of zoned plagioclase (up to 4 mm).

### Intrusive rocks

There is a close spatial relationship between basic intrusive rocks and the northern group of extrusive rocks. The intrusive rocks are predominantly olivine and hornblende dolerites, but picrites/peridotites are also present. They form substantial sill-like bodies, that are particularly prominent in the southern part of the Saltash division, and in the area between Menheniot and Liskeard. There a westwards decrease in extrusives is accompanied by an increase in intrusions.

A typical hornblende dolerite forms the prominent Berry Hill [374 587], south of Landrake. Phenocrysts of hornblende, augite and Fe-oxide (up to 4.0 mm across) are contained in subophitic and granular intergrowths of plagioclase, clinopyroxene and apatite. Hornblende appears to be a late phase, replacing and mantling augite; biotite also occurs as a late phase in the groundmass. The primary ferromagnesian minerals are replaced by chlorite, plagioclase by white mica, epidote by pumpellyite, and Fe-oxide by sphene (E 66413).

Formerly described under the name proterobase (Ussher, 1907), a suite of biotite-bearing potassic dolerite intrusions occur on the banks of the Lynher near Grove [379 575], at Ernesettle [4395 5911 to 4391 5900] and in a quarry below the Tamar road bridge [4375 5869]. Ranging from coarse-grained (up to 10 mm) subophitic (E 62271-2) to fine-grained granular dolerite (E 62270), they contain primary hornblende, considerable amounts of biotite and apatite, in a generally panidiomorphic feldspar groundmass.

Ultrabasic rocks are subordinate to the dolerites but form substantial bodies at Clicker Tor [285 614] and Broadmoor [3952 5814] and a thinner sill at Criffle Mill [329 607]. The Clicker Tor intrusion, described by Teall (1888) as an augite-picrite, comprises 60 per cent serpentinised olivine, 30 per cent poikilitic and interstitial augite and subordinate hornblende (E 67668; Barton, 1994). The body at Broadmoor consists of a variably serpentinised peridote with primary fresh olivine, clinopyroxene, brown amphibole, biotite, apatite, feldspar and magnetite. Olivine and serpentine pseudomorphs are poikilitically enclosed in large crystals (up to 10 mm) of clinopyroxene. Late magmatic alkali amphibole partially replaces the pyroxene, and late biotite and alkali feldspar are interstitial. A green mafic phyllosilicate (possibly corrensite) replaces much of the groundmass and some of the biotite, a probable result of low-grade regional metamorphism. The texture and mineralogy of this rock (E 66415) resemble the hydrous ultramafic and mafic rocks of the Polyphant Complex, west of Launceston, and possibly represents a cumulate from the basalt magma.

## IGNEOUS ROCKS IN THE DINANTIAN

**Extrusive rocks** occur immediately to the north of this area, at Newbridge [347 676] in the Tavistock district (Sheet 337) where thick pillow basalt is associated with chert of the Newton Chert Formation. Representatives of that volcanic sequence are present in the Plymouth district, but apart from the basaltic lava shown within the Newton Chert near Burcombe [406 666] crops are insufficently large to be represented on the 1:50 000 Series map.

**Intrusive rocks** within the Dinantian are also sparse but a general close association of the Brendon Formation and basic sills recorded in the Tavistock district extends into this district. On Viverdon Down [384 670] a thick gabbroic sill within the Brendon Formation forms the hanging wall of the topmost thrust sheet of the outlier. At Pentillie Park Farm [350 671] an ultramafic intrusion transgresses the Brendon Formation.

The basic igneous rocks of the area invariably show alteration of their primary mineralogy due to regional metamorphism in the sub-greenschist to zeolite facies. The submarine lava has been extensively carbonated, the activity of $CO_2$ suppressing the development of diagnostic hydrous calc-silicate assemblages. A typical secondary assemblage is calcite+albite+chlorite+sphene, with the replacement of the primary plagioclase and ferromagnesian minerals and the groundmass.

## PERMIAN IGNEOUS ROCKS

Emplacement of the Cornubian granite batholith accompanied the regional extension that immediately followed Variscan compressional deformations and metamorphism (Holder and Leveridge, 1994). It is a composite body; U–Pb monazite ages for the individual major plutons span an interval of ~20 Ma in the Early Permian. The Carnmenellis Granite is the oldest (293.7 ± 0.6 Ma, Chesley et al., 1993; 293.1 ± 1.3 Ma, Chen et al., 1993) and the Land's End Granite is the youngest (274.5 ± 1.4 Ma, Chen et al., 1993; 274.8 ± 0.5 Ma, Chesley et al., 1993). The Bodmin Moor Granite (291.4 ± 0.8 Ma, Chesley et al., 1993) lies just off the north-west corner of the district and the Dartmoor Granite (280.4 ± 1.2 Ma, Chesley et al., 1993; 281.0 ± 0.8 Ma and 285.3 ± 0.8 Ma for megacrystic and weakly megacrystic facies respectively, Chen et al., 1993) is to the north-east of the district. The Kingsand Rhyolite Formation and associated elvan dykes are probable surface and hypabyssal expressions of the Bodmin intrusive episode (see chapter 6).

Chemical analysis of the **lamprophyre dykes** of the region indicate derivation from the mantle. They have been dated by the K–Ar method at 296 ± 2.5 Ma (Rundle, 1980). If all the dykes are of similar age they were emplaced at an early stage of regional extension and at about the same time as the earliest cupolas. In the Plymouth area a lamprophyre dyke crops sporadically between Chapel Farm [417 659] and south-east of Furzehill Bridge [4417 6617]. On the east bank of the River Tamar, north-east of North Hooe [4229 6594], the dyke is broken up into a series of veins with intervening screens of slate. Individual veins are up to 1.5 m across in a zone of lamprophyre and slate screens 30 m in width. The rock is a minette consisting mainly of biotite, much of which has been altered to chlorite, potassium feldspar, Fe-oxide and corroded quartz xenocrysts up to 3 mm across (E 56573). Granite xenoliths, up to 70 mm across, are common and are enclosed by flow-aligned chlorite and biotite in the lamprophyre (E 54139). The granite is a non-megacrystic, medium-grained granite showing no preferred orientation of mineral grains.

**Elvan** dykes are associated with the Kingsand Rhyolite Formation, and similar quartz-feldspar-porphyry dykes, trending approximately east–west, occur in the northern part of the district. The most northerly of these forms a discontinuous crop from Down Wood [4477 6533] eastwards out of the area. Along its length phenocryst content, grain size and alteration are variable. It is coarsely phenocrystic near Lopwell [475 649], coarse-grained on Roborough Common [510 650], fine-grained to the west at Hole Farm [4610 6517] and heavily sericitised and chloritised near Southway [4828 6177 to 4865 6185].

## IGNEOUS ROCK COMPOSITION AND TECTONIC DISCRIMINATION

The Devonian to Carboniferous basic extrusives and intrusions of the district are predominantly within-plate alkaline basaltic rocks (Figure 26a), similar to those elsewhere in this belt of the province (Floyd, 1982, 1984; Rice-Birchall and Floyd, 1988). A mantle source for the Upper Devonian basalts, sampled from just south of Saltash, is indicated by the $\Sigma_{Nd}(T)$ values, and the Th/Ta and La/Ta abundances (Merriman et al., 2000). These isotope and geochemical characteristics suggest that the basalts were derived from Ocean Island Basalt (OIB)-type magma. Such a source is consistent with the low La/Nb and Th/Nb ratios shown by these rocks. The Lower and Middle Devonian and Lower Carboniferous rocks have similar ratios, the result of Nb and Ta enrichment during genesis of OIB (Weaver, 1991). Eruption of unmodified OIB melts generated in the mantle is an indicator of strong extension (McKenzie and Bickle, 1988) through the Devonian and into the Carboniferous. A minor proportion of the rocks, particularly within the Lower Devonian, have calc-alkaline affinities (Figure 26b). This same category has been recognised in the Lower Devonian of south Devon and attributed by Merriman et al. (2000) to crustal contamination of mantle-derived basaltic magmas during the early stages of crustal rifting.

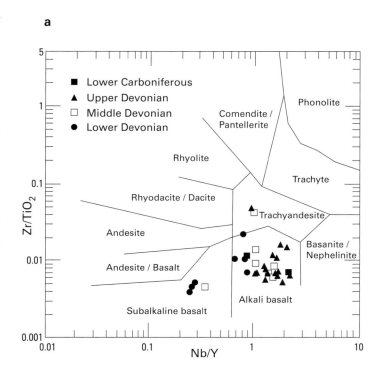

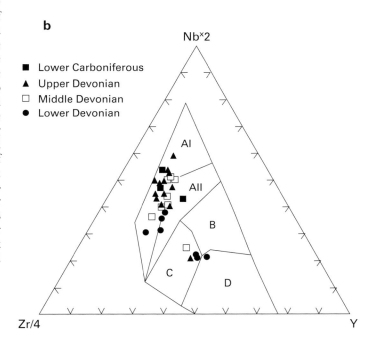

**Figure 26** Classification and discrimination of Lower Devonian to Lower Carboniferous basaltic rocks based on geochemical data.

a.  Data plotted on the discriminant diagram of Winchester and Floyd (1977)
b.  Data plotted on the discriminant diagram of Merschede (1986)

Al    Within plate alkali-basalt
AII   Within plate alkali-basalt + within plate tholeiitic basalt
B     P-type MORB
C     Within plate tholeiitic basalt + volcanic arc basalt
D     Volcanic arc basalt + N-type MORB

# NINE

# Metalliferous mineralisation

Classical models of metallisation in the province emphasise the role of mineralising fluids originating in, and dispersing from, the granite intrusions (Dewey, 1925; Davison, 1926; Dines, 1934) in the late Carboniferous/Early Permian. These take little account of the complex construction of the Variscan belt in south-west England and the major input to it of metals from hydrothermal events associated with Devonian and Carboniferous volcanism (compare with Scrivener and Bennett, 1980; Leake et al., 1985; Scrivener et al., 1989; Alderton, 1993). Recent studies (Shepherd and Scrivener, 1987; Scrivener et al., 1994) have shown that vein mineralisation was also associated with low temperature saline formation water circulation in the Triassic period. This extended history of mineralisation is particularly well demonstrated in the Plymouth district, which lies at the edge of the main zone of Sn–W–Cu–As deposits linked to granite emplacement.

The regional range of ore deposits, in generalised chronological order, can be summarised as follows:

| | |
|---|---|
| Post-granite | crosscourses |
| Granite-related | main-stage polymetallic veins |
| | tourmaline-quartz orebodies |
| | greisen-bordered veins |
| | pegmatites |
| | skarn deposits |
| Pre-granite | pre-granite veins |
| | syngenetic and epigenetic metal |
| | concentrations |

The Plymouth district has examples of epigenetic mineralisation relating to Early Devonian (iron) and early Carboniferous (manganese) basic volcanic rocks. Minor antimony-bearing veins probably represent the remobilisation of low-grade epigenetic metal concentrations in Middle Devonian tuffs and lavas, by regional metamorphic fluids during the later part of the Variscan earth movements. There is minor granite-related hydrothermal vein mineralisation at the north-western boundary of the district, and well-developed late, north–south-trending Pb–Ag–Zn–F veins in the Herodsfoot, Menheniot and Bere Alston areas.

## SYNGENETIC AND EPIGENETIC MINERALISATION

### Early Devonian

Silicification and associated sulphide mineralisation is present in basic volcanic rocks (lavas and tuffs) of the Bin Down Formation. This is well displayed at outcrop in a disused quarry at Bin Down [2757 5758], and in the Widegates Borehole [2748 5747]. In this quarry, locally pervasive silicification of the basic lava host is cut by abundant thin veins of quartz. The pervasive silicification is accompanied by scattered crystals and aggregates of pyrite throughout the rock, resulting in brown staining by oxidation on weathered surfaces. The Widegates Borehole established a sequence of interbedded sandstone, siltstone and mudstone and sporadic thin beds of tuff, overlying amygdaloidal basalt breccia, hyaloclastite and tuff. These volcanic rocks are pervasively silicified and cut by at least two generations of, mostly thin, quartz veins. The silicification is particularly intense in the upper 5 m of the volcanic sequence, and is accompanied throughout by disseminated grains and aggregates of pyrite. A generalised sequence for the mineralising events shown in this core is as follows:

i   Pervasive **silicification** and sulphide mineralisation
ii  Fracturing and emplacement of **quartz veins**
iii Fracturing and emplacement of **quartz, feldspar and carbonate veins**

**Silicification** has resulted in the replacement of lava and volcanic breccia by microcrystalline quartz. In the upper part of the mineralised core, the siliceous replacement is pervasive, with clearly recognisable volcanic fragments and features such as vesicles replaced with, and filled by secondary quartz. The microcrystalline quartz matrix carries a variable abundance of finely crystalline pyrite throughout. Island aggregates of more coarsely crystalline quartz are present in places, together with anastomosing quartz veinlets bordered locally by wisps of pyrite. Ghost textures throughout the core show little evidence of deformation, vesicles being essentially circular in all sections, though most of the coarsely crystalline quartz aggregates show undulose extinction. The silicification occurred at an early stage, clearly predating Variscan deformation events.

In places, the wisps of pyrite noted above thicken laterally into irregular, finely crystalline lensoid bands, or more coarsely crystalline aggregates of euhedral to subhedral pyrite intergrown with quartz (Plate 15). Elsewhere within the silicified host, pyrite occurs as scattered aggregates and single crystals. Some of the larger pyrite aggregates comprise bundles of bladed, in some cases slightly curved, crystals; strongly suggesting pyrite replacement of original marcasite or arsenopyrite. Arsenopyrite itself is present as rare, minute, ragged crystals within larger masses of finely crystalline pyrite, and marcasite occurs in a similar form, but within microcrystalline quartz. In places, there is considerable alteration of pyrite to earthy red hematite, commonly in intergrowth with chlorite.

The core throughout is cut by narrow **quartz veins**, mostly less than 10 mm thick, which disrupt and brecciate the siliceous host and associated bands and

**Plate 15** Silicified and mineralised basaltic volcanic breccia from the Bin Down Formation. Stringers, wisps and disseminated pyrite are cut by early cleavage-parallel quartz veins and later, cross-cutting, quartz-feldspar veins. Specimen from Widegates Borehole (GS523).

10 mm

aggregates of pyrite. Much of this quartz is colourless and relatively free from inclusions, with a fabric of interlocking grains showing undulose extinction.

A later generation of **quartz**, **feldspar** and **carbonate veins** carry generally minor amounts of pink K–feldspar and cream-coloured carbonate (probably ankerite). Both generations of quartz veins bear numerous small (20–30 micron) 2-phase inclusions comprising an aqueous fluid and a vapour bubble. Agitated movement of the vapour bubbles on exposure to high levels of illumination of thin sections indicates relatively high levels of carbon dioxide, suggesting that the homogenisation temperatures of these inclusions is likely to be in the range 200 to 300°C. In some of the sections examined (e.g. E 70459) the later generation of quartz veins carry small equant inclusions with a double meniscus indicating two liquid phases, probably aqueous fluid and liquid carbon dioxide. Present throughout all the vein quartz and siliceous replacement are cross-cutting curtains of minute secondary inclusions, representing healed fractures.

The mineralisaton of the Widegates Borehole and adjacent outcrops is considered to represent an epigenetic hydrothermal event modified by later multiphase fracturing and hydrothermal activity. While the initial silicification has not been dated by isotopic means, it is clear from textural evidence that it preceded deformaton. It is also apparent that the same mineralisation did not extend to the enclosing host sedimentary rocks. It would seem likely that the silicification and sulphide mineralisation took place as a result of hot spring activity within a pile of volcanic rocks soon after their extrusion in the Early Devonian, before the accumulation of superincumbent sediment. This event would certainly have taken place at relatively low temperatures, and may originally have involved deposition of amorphous or cryptocrystalline silica and sulphide phases such as marcasite.

Both sets of quartz veins were produced late in the Variscan deformation sequence, a feature exhibited in

the volcanic strata and, more particularly, in the sedimentary rock envelope. The fluid inclusion temperature range and high carbon dioxide content are typical of late-stage metamorphic fluids. It is probable that fracturing and quartz veining represent post-deformation, possibly late Carboniferous, hydrothermal events in the waning stages of the Variscan orogeny, rather than activity in the Early Permian related to the emplacement of the Cornubian granites.

### Mid to Late Devonian

Mine workings and trials for antimony have been recorded at a number of localities in the district (Dines, 1956; Hamilton Jenkin, 1967), all within the crop of the Saltash Formation. This association of antimony vein mineralisation with Mid to Late Devonian slaty mudstone, basic volcanic rocks and contemporaneous greenstone intrusions is reflected in the Wadebridge district of north Cornwall. In that area it has been demonstrated (Clayton et al., 1990) that the antimony veins are derived from the remobilisation of syngenetic or epigenetic metal concentrations by metamorphic hydrothermal fluids during the late stages of the Variscan orogeny. The apparent stratigraphical control of antimony mineralisation can be demonstrated with certainty in the Ivybridge area to the east of the district, where substantial stratiform mineralisation, comprising iron/calcium carbonates with pyrite and minor tetrahedrite, has been intersected in exploration boreholes drilled within a Mid to Late Devonian sequence of basic volcanic rocks (Leake et al., 1985).

### Early Carboniferous

Stratiform manganese ore deposits are present within the Newton Chert Formation (Viséan) near Pillaton, and small workings or trials are recorded at Pillaton Mine

**Plate 16** Hand specimen of laminar manganese ore from Tordown Mine, Pillaton. Pale grey chert laminae are replaced by brown, granular Mn-bearing carbonate and by pink rhodonite, locally altered to black Mn oxides (GS524).

(see Dines, 1956) and Tordown Mine [3589 6492]. Material from the latter locality comprises pale and dark grey chert, in irregular laminae up to 20 mm, with interbedded pale grey siltstone and dark grey mudstone. In places, the interbeds between chert laminae are composed of pink rhodonite and rhodochrosite that grade laterally and vertically into chert or siltstone (Plate 16). Elsewhere, the chert laminae are partially replaced by pale brown manganese-iron carbonate, disrupted and in some cases brecciated by hairline veinlets of quartz, pyrite and manganese oxide minerals. Pyrite is a common minor mineral phase in specks, irregular aggregates and stringers, and there are traces of galena. Joints and local small pockets throughout the ore are filled with white clay, probably kaolinite. Thin veins (less than 15 mm across) of quartz cut across the mineralised rock in places, generally normal to the chert laminae.

Throughout the mineralised chert there is considerable secondary alteration to manganese oxide minerals. Oxidation has occurred along joints and on broken surfaces, also as irregular masses connected by thin veinlets. In places there is partial oxide replacement of carbonate bands. The oxide minerals present are pyrolusite, wad, psilomelane and manganite.

These rocks reflect a complex metallogenic history, which may be summarised as follows:

i   Precipitation of manganese silicate and carbonate phases contemporary with the deposition of laminated chert
ii  Partial replacement of chert by manganese-iron carbonate, perhaps during diagenesis or as a result of epithermal fluid movements
iii Recrystallisation during regional metamorphism
iv  Fracturing and quartz veining, most probably during late stages of the Variscan orogeny
v   Fracturing and oxidation by percolating groundwaters

A similar style of mineralisation has been described from early Carboniferous strata elsewhere in south-west England. Both in the Chillaton district of west Devon (see Dines, 1956) and the Teign valley to the east of Dartmoor (see Edwards and Scrivener, 1999) Viséan chert beds have been worked for manganese. Although basic volcanic rocks, notably spilitic lavas, tuffs and dolerite intrusions, are associated with the cherts in these areas, there is a similar though less well-marked association between cherts and basic lavas in the Pillaton area as well.

The nature of this manganese mineralisation and the spatial association with contemporaneous basic volcanic activity suggests an origin in a syngenetic exhalative-type system (Scrivener et al., 1989). The origin of this mineralisation by the expulsion and cooling of hydrothermal fluids of predominantly meteoric origin, causing deposition at or near the sea floor, is consistent with field relations from this district. It is therefore suggested that these deposits were formed initially from the discharge of hot brines from vents onto the sea floor, resulting in the precipitation of manganese-rich mineral phases associated with the formation of banded chert beds. Movement of manganese-bearing fluids, mostly in small-scale fractures, through the developing pile of chert and associated sediments resulted in partial replacement of the more permeable host rocks by manganese-iron carbonate minerals on a bed by bed basis. The combination of syngenetic and epithermal processes, together with subsequent modification by the thermal and tectonic events of the Variscan orogeny, has resulted in the typically complex textures seen in these deposits.

## PRE-GRANITE HYDROTHERMAL VEINS

Evidence of antimony mineralisation in the district includes former trials and workings near Pillaton [3800 6258], known as Wheal Leigh (Hamilton Jenkin, 1967), nearby to the north of Villaton [3890 6267], and at Tredinnick Mine [360 595], near Tideford. Little is known of the structure of these deposits as there is no current access to in situ mineralisation. Ussher notes a north-north-east-trending lode on the 1897 County

Series field slip Cornwall 37 SE, some 150 m to the south of Leigh Farm, and in the memoir (Ussher, 1907) refers to an occurrence of antimony ore in a west-south-west-trending fault near Tredinnick.

Hamilton Jenkin (1967) noted the occurrence of stibnite at Wheal Leigh and Tredinnick Mine, and Dines (1956) recorded the presence of quartz and calcite veinstone in dump material at the latter site. A specimen collected at Tredinnick Mine (E67345) consists of a fine breccia of clasts of dark grey slaty mudstone cemented by overgrown quartz, sulphides and sulphosalts (Perez-Alvarez, 1993). Using reflected light microscopy, four ore mineral phases can be recognised, namely stibnite and boulangerite, minor sphalerite and rare jamesonite. Electron microprobe analysis demonstrated that the sphalerite contains between 1.1 and 2.2 wt% Fe, a range that is consistent with an origin in a hydrothermal vein system.

Quartz and quartz-carbonate-feldspar veins have been ascribed above to fracturing and the movement of hydrothermal fluids during the late stages of the Variscan orogeny. In the case of the Bin Down Formation, the presence of fluid inclusions with an estimated temperature homogenisation ($T_h$) in the range 200 to 300°C and high levels of $CO_2$ suggest the involvement of low-grade regional metamorphic fluids. Fluids of this type have contributed to metalliferous mineralisation elsewhere in Cornwall, such as the antimony deposits of the Wadebridge district in north Cornwall, which include numerous small workings in scattered veins of roughly north–south trend. The minerals worked from these deposits include jamesonite and bournonite, with some galena, sphalerite, chalcopyrite and pyrite, in a gangue of quartz and siderite. While some authors (Hosking, 1964; Edwards, 1976) consider this mineralisation to be typical of the low-temperature crosscourse veins found elsewhere in Cornwall, recent fluid inclusion studies (Clayton et al., 1990) have shown that the antimony veins were formed from fluids at higher temperatures, with $T_h$ in the range 280° to 315°C. These inclusion fluids are of low salinity, based on NaCl, with abundant non-aqueous volatiles ($CO_2$ and $CH_4$). Such brines are typical of low-grade regional metamorphic fluids. These data, together with the structural setting of the veins in north-north-east–south-south-west shear zones, suggest that the antimony mineralisation of north Cornwall was effected during the late stages of Variscan deformation. A similar origin is suggested for the antimony deposits of the Plymouth district.

## GRANITE-RELATED HYDROTHERMAL VEINS

The east–west-trending 'mainstage' Sn–W–Cu–As veins of south-west England are spatially related to, and in some cases fall within, the crops of the various components of the Cornubian granite batholith. Immediately to the north of the district, extending eastwards from the southern part of the Bodmin Moor Granite, are the major polymetallic deposits of Caradon Hill and the Phoenix mines (Sn–Cu) and, farther east still, the mines

of the Kit Hill and Gunnislake areas (Sn–W–Cu–As). Fluid inclusion and isotope studies (Shepherd et al., 1985; Alderton, 1993) clearly demonstrate the genesis of the mainstage veins initially from saline hydrothermal fluids of direct magmatic (i.e. granite) departure, with subsequent dilution by circulating brines of meteoric origin.

Vein mineralisation of the granite-related type is sparsely represented in the district, the only significant example being at Kilham Mine [2050 6695], where a polymetallic east–west-trending vein bearing a complex assemblage of cassiterite, with sulphides of copper, arsenic, iron, zinc and lead, was worked for tin. While there are no fluid inclusion data for Kilham Mine, quartz from a nearby east–west-trending sphalerite-chalcopyrite vein at Gill Mine [2940 6795], in the adjoining district to the north, yielded $T_h$ values in the range 168 to 270°C and salinities of 1.5 to 3.5 wt% NaCl equivalents (Alderton, 1978). The data for Gill Mine compare with values quoted by Bull (1982) for granite-related poly-metallic vein mineralisation in the Gunnislake area of the Tavistock district.

## CROSSCOURSE MINERALISATION

The district lies at the southern margin of the Cornubian metalliferous province, and contrasts with the more important metal mining districts elsewhere in Devon and Cornwall, where much of the production of tin, copper and arsenic was derived from veins of roughly east–west trend. In the Plymouth district, north–south-trending hydrothermal vein complexes ('crosscourses') were notable for their production of lead, silver, zinc and fluorspar. Within the district are three centres for this type of mineralisation, namely the Herodsfoot, Menheniot and Bere Alston (Tamar valley) areas.

A north–south-trending vein system lying to the west of the village of Herodsfoot has been exploited over a strike length of more than 1300 m. The vein filling was stated by Dines (1956) to be chiefly saccharoidal and chalcedonic quartz with siderite and pyrite and strings, aggregates and bunches of argentiferous galena present throughout. Sphalerite, baryte and dolomite are present in vughs. Bournonite, antimonite and chalcopyrite have also been noted. At one time, the mine was noted for spectacular specimens of tetrahedrite, a minor sulphosalt mineral. The secondary minerals cerussite and pyromorphite were recoded from the veins at shallow depths. Fluorite has not been recorded from the Herodsfoot veins.

The Menheniot lead-silver mineralisation comprises two north–south-trending vein systems extending northwards from Menheniot. Whilst no underground workings could be examined during this survey, some information on the Menheniot lode mineralogy and paragenesis is given by Foster (1878) and by Dines (1956). At Wheal Mary Ann the principal vein filling is quartz, with banded chalcedonic material commonly present at the lode walls. This is overgrown by crystalline quartz with galena, and then by siderite. At least two generations of fluorite may be present predating and postdating the quartz-galena stage. Late-stage minerals include calcite, siderite and pyrite, with baryte locally present as vugh fillings. Fluorite is a major gangue mineral at Wheal Mary Ann and has been worked from the dumps.

At Wheal Trelawny, arsenopyrite and pyrite were present in workable quantities, together with argentiferous galena; minor chalcopyrite is present throughout, becoming more abundant at depth. The gangue minerals are quartz, calcite and yellow fluorite, with minor baryte; the secondary lead minerals cerussite and pyromorphite were recorded at shallow depth.

The lodes of Wheal Wrey and Wheal Ludcott at shallow depth comprise quartz, argillised inclusions of slate country rock and limonite, with pyrite, chalcopyrite, sphalerite and galena. At greater depth the vein fillings are of chalcedonic quartz and calcite with a higher proportion of galena. Fluorite is present in some quantity at Wheal Wrey, and as a minor phase at Wheal Ludcott. Large late-stage crystals of baryte were recorded from Wheal Wrey. The veins of both mines are intersected and displaced by east–west fractures dipping about 60° southwards. At certain of the intersections, bunches of silver minerals, notably argentite, pyrargyrite, native silver, proustite and stephanite were recorded in association with siderite.

The crosscourse veins of the Tamar valley extend southwards from Calstock, on the adjacent district to the north, for a distance of about 6 km, and comprise two systems lying about 1 km apart, cutting folded and faulted slaty mudstone, shale and sandstone of Late Devonian and early Carboniferous age. These veins occupy fractures, mostly dipping steeply towards the east, filled with quartz and brecciated host rock in which bodies of galena, sphalerite and fluorite are sporadically present (Dines,1956). There is no access to the workings at the present day and very little detailed information concerning the nature of the lodes in the literature. A paragenetic study of the Tamar valley lodes by Bull (1982) states that fluorite and sphalerite are deposited at an early stage, followed by galena. Both granular and chalcedonic quartz are common, with the chalcedonic material generally later in the paragenetic sequence. Minor amounts of calcite and siderite are present together with tetrahedrite, chalcopyrite, arsenopyrite, pyrargyrite, pyrite and marcasite. The Tamar valley lodes are cut and displaced in places by southward-dipping east-west fracture zones termed 'slides'.

Crosscourse vein mineralisation of the district is, in terms of mineralogy and paragenesis, typical of low-temperature lead-zinc-silver-fluorite ore deposits throughout much of Europe. Over much of south-west England, the relatively great distance of the lead-zinc veins from centres of granite-related tin-copper mineralisation led to speculation that the lead-bearing veins were deposited as an outer zone of the granite-related metallisation (e.g. Dines, 1956). In certain cases outside the district this appears to be true, for example the veins of the Perranzabuloe–Newlyn lead mining area of west Cornwall were formed from low salinity, sodium chloride-dominated brines (Alderton, 1978) similar in composition and temperature range to the fluids responsible for the mainstage tin-copper ores (Bull, 1982; Scrivener et al., 1986).

For the Plymouth district, fluid inclusion studies (Alderton, 1978; Bull, 1982; Shepherd and Scrivener, 1987) have demonstrated that the Menheniot and Tamar valley crosscourse veins were deposited from low temperature fluids of high salinity and of distinctive chemistry, with high levels of calcium chloride. A microthermometric study by Alderton (1978) included data from fluorites from the Menheniot veins, and provided a range of $T_h$ values of 112 to 172°C, and of gross salinities 23.4 to 24.8 wt% NaCl eqivalents. In Alderton's study the relative abundances of sodium and calcium chlorides were not determined. The work of Shepherd and Scrivener (1987) on inclusion fluids in fluorites from both the east and west Tamar valley vein systems, also demonstrate a very restricted range of inclusion fluid salinities and of $T_h$. In this case the $T_h$ range was 110 to 170°C and the salinity range 19 to 27 wt% NaCl equivalents. It was further indicated, on the basis of eutectic melting determinations that the fluids contain the equivalent of 11 to 15 wt% NaCl and 9 to 13 wt% $CaCl_2$.

Using the Tamar valley microthermometry together with data from fluorite Rare Earth Element analyses, Shepherd and Scrivener [1987] concluded that the low-temperature fluids of the crosscourse veins were highly evolved formational brines which had originated in, and been expelled from, nearby sedimentary basins. The very restricted range of $T_h$ and salinity values demonstrate a lack of fluid mixing in contrast to the complex genetic history perceived in granite-related vein systems. The role of basinal brines in late-stage low-temperature mineralisation is not restricted to the Plymouth district. Scrivener et al. (1994) have drawn attention to similar fluids in crosscourse systems in west Cornwall and in the Teign valley of Devon, also in gold-carbonate veins at Hope's Nose, Torquay, and in manganese replacement mineralisation in Permian red beds near Exeter.

## CHRONOLOGY OF MINERALISATION

Isotopic dating of ore deposits from south-west England, for the most part, has been targeted at understanding the age of mainstage Sn–Cu–As–W mineralisation in relation to the emplacement of the Cornubian granites. The most thorough investigations of emplacement ages of the exposed components of the Cornubian batholith are by Darbyshire and Shepherd (1985 and 1987) using Rb–Sr techniques, and by Chesley et al. (1993) and Chen et al. (1993) using a combination of U–Pb and Ar–Ar methods. These give a range of ages from 298 to 270 Ma for the major and minor plutons and for the satellite quartz-porphyry dykes ('elvans').

Halliday (1980) reported K–Ar and Rb–Sr mineral ages for a range of granite-related ore deposits in south-west Cornwall. Metal-bearing pegmatites and greisens yielded ages of about 285 Ma, while feldspars from mainstage structures gave a mean age of 270 Ma. The same feldspars on K–Ar analysis gave a range of ages from 212 to 244 Ma, and the author speculated that the inferred argon loss was due to interaction of the vein minerals with later generations of hydrothermal fluid. Work by Bray and Spooner (1983) on kaolinised areas of the St Austell Granite included a substantial number of K–Ar analyses, which indicated a mean age of 280 Ma for sheeted vein mineralisation and associated kaolinisation, together with a cooling age of about 250°C for the granite at 274 Ma. In addition to the granite emplacement ages, Darbyshire and Shepherd (1985) also give a fluid inclusion Rb–Sr isochron age of 269 ± 4 Ma for the mainstage tin mineralisation at South Crofty Mine in west Cornwall. The lack of data for the latter, sulphide-rich stages of the mainstage veins prompted Chesley et al. (1991) to attempt Sm–Nd dating of fluorite from those structures. Isochron ages of 259 ± 7 Ma and 266 ± 3 Ma, respectively, were obtained from suites of fluorite samples from South Crofty Mine

and Wheal Jane in west Cornwall. Chesley et al. (1993) demonstrated a distinct mineralisation history for each granite pluton starting with greisen and tourmaline veins coeval with magmatic emplacement and continuing for 3 to 4 Ma, followed by protracted fluid circulation and the formation of mainstage sulphide veins for up to 25 Ma.

De la Beche (1839) to Dines (1956) and Hosking (1964) all record crosscourses as cutting, and therefore postdating, the mainstage veins. While the mainstage mineralisation has commonly been regarded as directly related to the emplacement of the Cornubian Batholith, or to thermal anomalies around that body, the age of the crosscourses remained a matter of speculation. De la Beche (1839), noting the presence of north–south faults affecting the Cretaceous and Cainozoic strata of the Blackdown Hills of east Devon, suggested that the crosscourse mineralisation might be of 'Tertiary' age. Support for Mesozoic and Cainozoic ages was given by Collins (1912) and Hill and MacAllister (1906), respectively, although no conclusive evidence was cited. More recently, Durrance et al. (1982) have suggested that crosscourse mineralisation was due to 'an influx of Mesozoic sea water', a view which conflicts with the inclusion fluid data and conclusions quoted above.

Until recently, radiometric dating of crosscourse mineralisation has been restricted to U–Pb and Pb–Pb determinations. Moorbath (1962) obtained ages in the range 280 ± 20 Ma for a group of galena specimens from south-west England which included material from the Menheniot and Bere Alston veins. Uranium minerals from a number of localities in Devon and Cornwall were analysed by Pockley (1964) and Darnley et al. (1965). The results of these studies tentatively suggested periods of uranium mineralisation at 290 Ma, about 225 Ma and 50 to 60 Ma.

During the course of the survey, Scrivener et al. (1994) applied Rb–Sr isotope analysis to fluorite and inclusion fluids in quartz (compare Darbyshire and Shepherd, 1985) from the Tamar valley crosscourse veins. The eastern vein system of the orefield was sampled at Buttspill [437 678], Lockridge [439 665], Furzehill [437 656], and South Tamar Consols [437 645] mines, representing a north–south distance of about 3 km. The data obtained yielded an isochron age of 236 ± 3 Ma (Figure 27), which suggests a mid to late Triassic age for the Tamar valley crosscourse system.

The age of the Menheniot crosscourse mineralisation was investigated by Sm–Nd isotope

| Age | Mineralisation type | Metals | Locality/host |
|---|---|---|---|
| Mid-Triassic 230–240 Ma | Hydrothermal vein (Post-granite crosscourse) | Pb–Ag–Zn–Cu–As–F | Tamar valley |
| | | Pb–Ag–Zn–Cu–As–F–Ba | Menheniot |
| | | Pb–Ag–Zn–Cu | Herodsfoot |
| Early Permian 270–290 Ma | Hydrothermal vein (granite related) | Sn–As–Cu–Zn | Kilham Mine |
| Late Carboniferous Stephanian 290–300 Ma | Hydrothermal vein (Pre-granite) | Sb–Pb–Zn | Saltash Formation |
| Early Carboniferous Viséan 332–350 Ma | Stratiform exhalative | Mn | Newton Chert Formation |
| | *Sedimentary diagenetic* | *Pb–Zn* | *Egloskerry* |
| Mid to Late Devonian Givetian–Frasnian 367–381 Ma | *Stratiform exhalative/replacement Volcanic association* | *Fe–Sb* | *Ivybridge* |
| | *Stratiform Basic igneous association* | *Sb–As–Pb–Cu* | *Portquin (N Cornwall)* |
| Early Devonian Pragian 390–396 Ma | Stratabound epigenetic association with basic volcanics | Fe–As | Bin Down Formation |

**Table 14** Summary of metalliferous mineralisation events in the Plymouth district and surrounding area (in italics).

analysis of fluorite specimens collected from the spoil dumps at Wheal Wrey in the eastern vein system, and from Wheal Hony, Wheal Trehane, Wheal Trelawney and Wheal Mary Ann in the western complex (Darbyshire, 1995). The data yielded an isochron age of 230 ± 10 Ma (Figure 27), the median of which places the Menheniot mineralisation in the early part of the Late Triassic, with the extremities of the error in the Norian and Anisian stages of the Triassic (Harland et al., 1989).

The significance of a Triassic age for the crosscourse mineralisation was considered by Scrivener et al. (1994) who concluded:

i The highly saline, calcium chloride-rich brines originated within Permo-Triassic sedimentary basins flanking, and to some extent covering the Variscan massif

ii The restricted composition range of the crosscourse brines suggests that they represent an equilibrium state deep within the Permo-Triassic red-bed sequence and that any near-surface groundwater influx was minimal

iii Movement of the formation brines was in response to extension-driven fault-controlled subsidence; north-south fracturing occurred during the middle part of the Triassic, initiating density driven circulation of the brines in the Variscan basement

iv The presence in the crosscourse systems of economic minerals is governed by the proximity of appropriate metal reservoirs, such as the mineralised volcanic-sedimentary sequences of the Middle/Late Devonian and early Carboniferous basement

It is thus possible, within the Plymouth district, to trace the metallogenic history from the Early Devonian to the Late Triassic (Table 14) in a variety of differing geological environments and tectonic settings. It is a feature of the south-west England Variscan province that most of the metallogenic events took place at relatively shallow depths during periods of crustal extension.

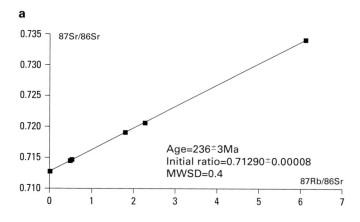

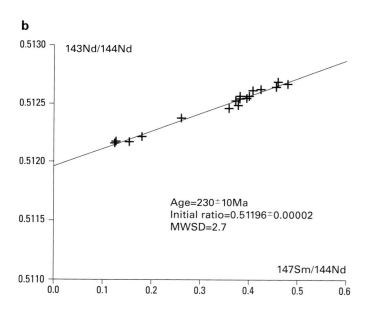

**Figure 27** Radiometric ages of crosscourse veins.

# TEN

# Variscan deformation and metamorphism

## STRUCTURAL GEOLOGY

Variscan folds and cleavage were observed, the Cawsand and Portwrinkle faults deduced from displacements of mapped divisions, and the presence of thrusting was noted by Ussher (1907) during the first survey. The regional significance of the Cawsand, Portwrinkle and Portnadler faults was subsequently recognised by Dearman (1963) who described them as dextral wrench-faults ascribing them to mid-Tertiary deformation, a view with which Lane (1970) concurred. Plunge variation of folds in Cornwall was first recorded at Britain Point [352 358] near Portwrinkle by Dearman (1964), a feature subsequently recognised elsewhere in Cornwall as characterising zones of strong differential compression (Dearman, 1969), and later more generally in shear regimes ('sheath folds' of Cobbold and Quinquis, 1980). The presence of thrusting in the Plymouth area was confirmed by Chapman (1983) and Chapman et al. (1984), who correlated major structures with those recognised in south Devon by Coward and McClay (1983). Seago and Chapman (1988) extended the concept to all the rocks of The Sound and lower Tamar valley with juxtaposition of their stratigraphical divisions interpreted to be the result of thrust dissection of a layered sequence. They also identified a confrontation of structural facing in the northern part of the Plymouth district.

During the resurvey polyphase deformation structures have been recorded in the district, largely related to continental collision and subsequent rebound in the Variscan Orogeny (Holder and Leveridge, 1994). The major episodes are two compressive deformation events $D_1$ and $D_2$, and a subsequent extension accommodated mainly by faulting. $D_1$ and $D_2$ were non-coaxial but sufficiently close for some $D_1$ structures to be enhanced/reactivated during $D_2$. In the case of thrusting, the most important element of the structural framework, attribution to episode is not always feasible and thrusts of both episodes are described together below.

There are three distinct major structural domains in the district differentiated largely on $D_1$ characteristics but in part modified and constrained by $D_2$ and later structures. They are the **Southern Domain**, that includes the rocks of the Looe Basin, South Devon Basin, and the southern part of the Tavy Basin, the **Northern Domain**, constituting the major part of the Tavy Basin rocks, and the **Culm Domain**, comprising the Carboniferous rocks of the St Mellion outlier.

### Southern Domain

FIRST MAJOR COMPRESSIVE DEFORMATION, $D_1$

Structures of $D_1$ are an ubiquitous cleavage ($S_1$), folds ($F_1$) on mesoscopic (minor, hand specimen to cliff size) and macroscopic (major, affecting mapped distribution) scales, thrusts and strike-slip faults.

*$S_1$ cleavage*

In all areas $S_1$ is an axial plane cleavage to folds of the $D_1$ deformation. $S_1$ is a slaty cleavage present in all of the finer grained argillaceous rocks. It is a penetrative pressure-solution cleavage in the coarser rocks of the Dartmouth Group, apart from the quartzites in which strained annealed grains show a secondary preferred dimensional orientation parallel to spaced pressure solution laminae [E 67234]. Cleavage is generally more widely spaced in the stratigraphically higher sandstone units. Volcanic rocks, and dolerite dykes are commonly cleaved, but the cores of some of the thicker sills are not. In the southernmost parts of the district, the preferred dimensional orientation of minerals of slaty rocks imparts a grain that is apparent locally on cleavage surfaces. Where recorded between the Rame and Portnadler faults it is parallel to the long axes of clasts, burrow fills, and aggregates of chlorite, extended in the plane of cleavage (Figure 28, d). The extension lineation there plunges moderately to the south-south-east. The internal reorganisation and recrystallisation producing the cleavage fabric, as in the other domains, constituted the main regional metamorphism (see p.109).

$S_1$ has a modal east–west strike west of the Portnadler Fault and to the east of the Portwrinkle Fault and its southern splay, the Rame Fault. Between those major faults cleavage typically strikes in the north-east quadrant (Figure 28a), but in the Portwrinkle area, between the Portwrinkle Fault and the Hoodney Cove Fault [3521 5387] some 700 m to the west, cleavage strikes north-west–south-east (Figure 28, a; north-east quadrant). Cleavage generally dips southwards moderately to steeply but there are notable areal variations within this broad pattern. In the Landrake area [36 61] cleavage is modally near-vertical and northward-dipping cleavage is common (Figure 28, b). Cleavage in the southern extremities of the area, about Rame Head and Polperro, whilst maintaining a similar geometrical relationship to bedding as to the north, dips steeply northwards (bedding is overturned and youngs southwards). In sporadic east–west zones a few hundred metres wide, as near Ince [395 564] and at St John [410 540], cleavage is variably steeply to gently northerly dipping. These areas of northerly dipping cleavage and related southerly facing folds are the result of later southerly vergent folding (see p.103).

*$F_1$ folds*

First phase folds of differing magnitude in the domain generally trend approximately east–west (e.g. Figure 28, c) in those areas where the axial planar $S_1$ has an approximately east–west strike. Between the Portwrinkle/Rame

**Figure 28** Stereographic (Schmidt) projections of structures in the Southern and Northern domains.

a.  Southern Domain $S_1$ cleavage on 1:10 000 Sheet SX 35 SE (south-west of the Rame Fault)

b.  Southern Domain $S_1$ cleavage on 1:10 000 Sheet SX 36 SE (south)

c.  Southern Domain $F_1$ axes (solid squares) and bedding/S1 intersection lineations (open triangles) on 1:10 000 Sheet SX 45 NW

d.  Southern Domain $F_1$ axes (solid squares), bedding/$S_1$ intersection lineations (open triangles) and $D_1$ extension lineations (half solid circles) on 1:10 000 Sheet SX 35 SE

e.  Southern Domain $F_2$ axes (solid squares) and $S_2$ cleavage (open squares); data from 1:10 000 sheets SX 35 SE and SX 44 NW southwest of the Rame Fault ($F_2$ symbols ticked) and SX 45 SE, SX 45 SW, SX 45 NW and SX 46 SW (south) to the north-west of the Rame Fault

f.  Southern Domain $S_1$ cleavage south of Liskeard on SX 25 NW (north) and SX 26 SW (south)

g.  Southern Domain $S_2$ cleavage south of Liskeard on SX 25 NW (north) and SX 26 SW (south)

h.  Southern and Northern domains $F_3$ axes (solid squares) and $S_3$ cleavage (open squares); ticked symbols are in the Northern Domain

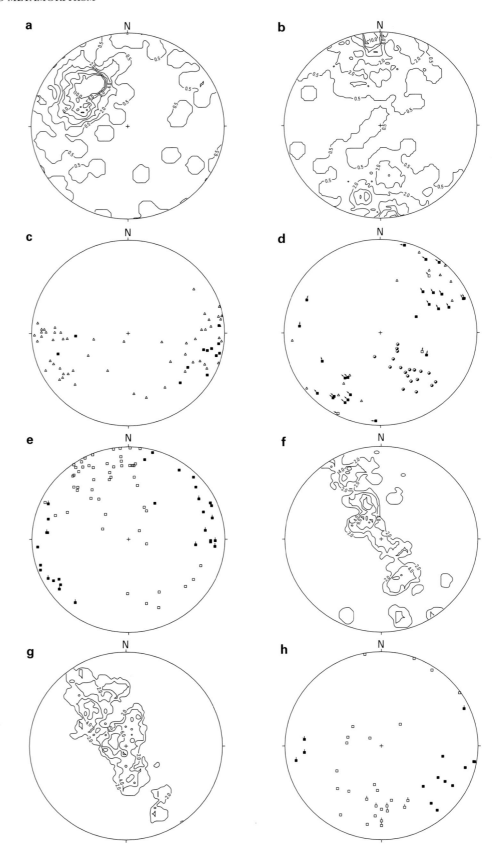

Fault and the Portnadler Fault, where $S_1$ has an approximately east-north-east–west-south-west strike, $F_1$ fold trend shows a similar relative sinistral rotation (Figure 28, d). Most structures are moderately to steeply inclined southwards with gentle to moderate plunge, but a few folds are reclined with high pitch angles. Minor folds are tight and higher order structures are tight to close. Hinge zones are rounded and show hinge zone thickening. Mapped major folds have a general trend between east-north-east and east-south-east but incipient sheath folding is present in mesoscopic structures. In the Lower Devonian rocks recorded fold axes locally show a wide variation in trend (Figure 28, d). A pitch variation of 60° is recorded near Britain Point [35150 53840] in near-vertical minor folds of quartzitic sandstone. Two closely related localities [3982 5185 and 3987 5181] near Freathy show folds plunging gently north-eastwards and reclined folds plunging moderately east-south-eastwards with axial plane cleavage of similar orientation and attitude. To the north such incipient sheath folding is generally less evident but is recorded locally, as in the hanging wall of the thrust between the Torpoint Formation and Saltash Formation near Neal Point [435 612] and in the footwall limestone of the thrust bounding the Plymouth Limestone Formation to the south at Mount Batten [4874 5305]. Folds face and verge northwards (with related faults), except where there is sheath folding or secondary refolding.

The mesoscopic folds occur as isolated couplets, in small families or as parasitic folds to major asymmetrical structures (see Figure 30). The largest of the recorded folds is that involving the rocks on the east side of The Sound where mesoscopically folded, steeply inclined and overturned bedding in the northern limb of the major northward-verging Bovisand Antiform extends over some 2.5 km in dip section. Major folds, of smaller dimensions, are mapped sporadically within sequences, as in the Devonport area, but most major folds are closely associated with the major thrusts bounding lithostratigraphical divisions. Major antiformal folds are recorded in the hanging walls of several of the thrusts, as at Looe [255 523] and Neal Point [435 612]. Extensive sections of the Whitsand Bay Formation in Whitsand Bay are largely devoid of folds but in the hanging walls of the thrusts bounding the nappes at Tregantle Cliff [387 526] and Portwrinkle [360 538] major $D_1$ antiformal culminations, subsequently extended, comprise a series of related macroscopic and mesoscopic overturned anticlines and synclines (Figure 29).

### Second major compressive deformation, $D_2$

$D_2$ structures, folds and cleavage, are developed sporadically through much of the succession, and are particularly intense to the south of Liskeard in the area north of the Staddon Formation and west of the Portwrinkle Fault where $D_1$ structures are locally transposed. Thrusting, on out-of-sequence thrusts or on reactivated earlier structures, is an important component of the $D_2$ deformation.

### $S_2$ cleavage

$S_2$ is an axial plane cleavage to $D_2$ folds. Second phase cleavage is a crenulation cleavage. Characteristically, it varies in intensity in fine-grained rocks from widely spaced crenulation of $S_1$ to cleavage banding transposing $S_1$ (E 70326). It is an irregularly spaced pressure solution and fracture cleavage in sandstone. Quartz veins are commonly associated with $D_2$ structures, mainly occupying dislocations along cleavage planes.

Modal strike of $S_2$ is east-north-east (Figure 28, e) and dips are generally steep to vertical but locally may be shallow to the south-south-east. In the St Keyne area, where $S_2$ is strongly developed and subsequent structures are not, cleavage dip is gentle to subhorizontal (Figure 28, g).

### $F_2$ folds

Second phase folds are small mesoscopic asymmetrical structures, generally open to close with rounded hinge zones, and upright to moderately inclined. Where $D_2$ structures are well developed, and $D_1$ structures strongly deformed (Figure 28, f), hinge angle and fold inclination (dip of $S_2$) decreases. Folds verge north-westwards and plunge gently north-eastwards or south-westwards (Figure 28, e). They are thus oblique to $D_1$ structures where those have an east–west general strike but trend parallel to them between the Portwrinkle and Portnadler faults. The rotation of $D_1$ structures between the faults preceded development of $D_2$ structures.

### $D_1$ AND $D_2$ FAULTS

The approximately coaxial nature of the main deformations in some areas means that differentiation of $D_1$ and $D_2$ thrusts is not always feasible. On the mesoscopic scale the absence of any association of secondary structures indicating $D_2$ movement points to the possibility of $D_1$ thrusting. The presence of associated secondary structures indicating $D_2$ movement does not preclude $D_1$ thrust movement. An additional complication is that subsequent extensional movements have affected most earlier faults, such that thrust related movement indicators may be disrupted or overprinted (Figure 29, b).

### $D_1$ thrusts

$D_1$ thrusts on the mesoscopic scale are exemplified in Tregantle Cliff and foreshore (Figure 30). There [3872 5262], small-scale thrust displacements on $S_1$ of less than one metre in folded sandstone beds are apparent (Figure 30, a). Thrust stacking associated with minor $F_1$ folding in a quartzitic sandstone packet [3874 5256] is evident (Figure 30, b); thrusts are parallel to cleavage where they penetrate the base of the packet, but are shallow and parallel to bedding within the packet. Minor, open, culmination anticlines are present where beds are repeated by overthrusting. Transport in the direction of vergence is to the north-north-west. In the Tregantle Cliff face [3882 5257] purple and green mudstone and siltstone of the Whitsand Bay Formation form the hanging wall of a sharply defined fault, moderately to gently inclined eastwards, with Bovisand Formation rocks forming the footwall (Figure 30, c). Minor $F_1$ folds in the hanging wall are in the overturned limb of a higher order structure, the core of which is some 20 m above the fault and about which $S_1$ is fanned. The fault is sub-

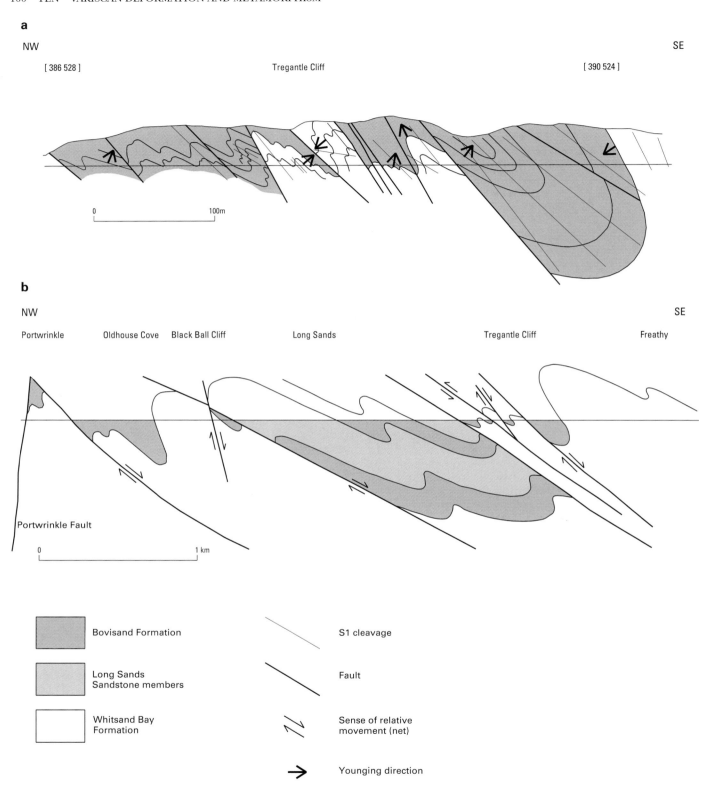

**Figure 29**   Profiles showing the relationship between the Whitsand Bay Formation and Bovisand Formation along the Whitsand Bay coastal section.

a. Tregantle Cliff: a structural profile showing the folded and faulted boundary between the formations.
b. A sketch of the coastal section between the Portwrinkle Fault and Freathy, based on recorded details, showing the main structures affecting distribution.

**Figure 30** Folding and thrusting at Tregantle Cliff.

a. Small-scale thrust displacement parallel to $S_1$ in folded sandstone beds (stippled), Tregantle Cliff foreshore.

b. $D_1$ thrust stacking of a quartzite sandstone packet in purple and green slaty mudstone, Tregantle Cliff foreshore.

c. $D_1$ overthrusting of Whitsand Bay Formation rocks (D) on to Bovisand Formation rocks (B), Tregantle Cliff.

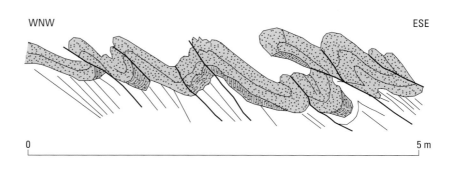

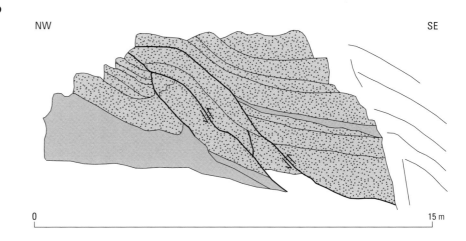

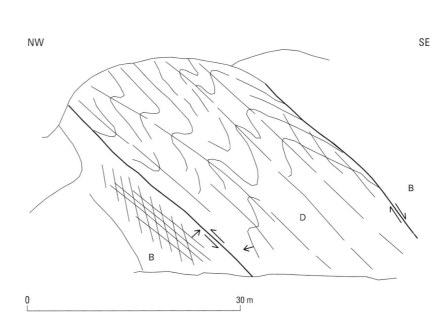

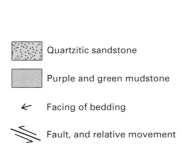

parallel to cleavage in the overturned limb, and is oblique to cleavage in the right-way-up limb and in the footwall where beds are the right way-up. The relationships are consistent with $D_1$ dislocation subparallel to the lower right-way-up limb of a large mesoscopic fold couplet, and with transport up dip.

Major thrusts (those with a lateral extent in excess of 1 km) are recognised and mapped as forming many of the lithostratigraphical boundaries of the district. Unlike the thrust forming the northern boundary of the Staddon Formation on the east side of The Sound (Plate 17) most are unexposed. Their presence is

**Plate 17** The Crownhill Thrust in the bank of the River Tavy [4499 6120] looking eastwards. In the 1 m of deformed and brecciated slates, early up-dip movement indicators are largely obscured by later down-dip extensional structures. The fault juxtaposes the Saltash Formation of the South Devon Basin in the hanging wall with the Torpoint and Tavy formations of the Tavy Basin in the footwall (GS525). Hammer is 0.30 m long.

indicated by the juxtaposion of rocks of differing age and character, the way-up of the adjacent sequences in the vicinity of the boundary, and the attitude and orientation of the boundary in relation to the structures within the juxtaposed sequences. Intraformational thrusts are recognised as major structures only where, as in the Landulph division of the Saltash Formation, similar rocks of differing age are juxtaposed [4361 6135]. The possible presence of intraformational thrusting in areas where detailed biostratigraphical evidence is lacking cannot be discounted.

### $D_2$ thrusts

$D_2$ thrusts, in a similar manner to the $D_1$ thrusts, are present on all scales. $D_2$ thrusts on the mesoscopic scale are, for example, present in the cliffs to the north-west of Ravens Cliff [4583 5258]. There, the junction between massive hyaloclastite and overlying grey slaty mudstone of the Saltash Formation (Jennycliff division) describes an open asymmetric fold couplet that verges northwards. $S_2$ steeply inclined southwards is an axial plane cleavage to the folds. Dislocation between the volcanic rock and mudstone on the gently inclined southern limb passes northwards over the anticlinal fold hinge into an imbricate set of thrusts through the northwardly inclined steeper limb. The faults are closely associated with small-scale open to tight gently inclined $F_2$ folds.

An example of an essentially $D_2$ major thrust bounds the rocks of the Torpoint Formation overlying Saltash Formation rocks in the Ernesettle/Whitleigh area. The trace of the fault in places [e.g. near Budshead Wood 461 597] contours around the topography, and a gentle dip to the south is indicated. Within the Saltash Formation of the footwall, a zone of thinly bedded limestone extends south-eastwards from Tamerton Lake [459 603]. About 1 km west of Crownhill, this zone is occluded by the overriding thrust sheet, to reappear approximately 0.5 km to the east, in the bottom of a valley, through an erosional window in the thrust

hanging wall (BGS, 1996b). An isolated klippe of the hanging wall Torpoint Formation is present at Whitleigh [475 598] some 0.5 km to the north of the main trace of the thrust. The fault cuts across both stratigraphy and $S_1$ indicating its post-$D_1$ age. A minimum transport distance of about 0.9 km is evident on the thrust.

### Strike-slip faults

The main oblique faults striking between north-west and north-north-west (Figure 1) are the Cawsand Fault, the Portwrinkle Fault (with its southern splay the Rame Fault) and the Portnadler Fault. These faults have dextral offsets of the order of 4 km, 7.5 km (with up to 2 km on the Rame Fault) and 1.5 km, respectively. Fault beccias are up to 2 m wide, as in the case of the Portwrinkle Fault [3594 5383] which comprises 0.5 m of clay gouge and 1.5 m of intensively sheared country rock. Rotation of $S_1$ and bedding over several tens of metres adjacent to the faults and drag on subordinate subparallel faults (up to 70 m from the main fault in the case of the Cawsand Fault at Kingsand [4355 5055]) indicate that the major movements have been dextral strike-slip. More localised steeply plunging folds of $S_1$ near the Cawsand Fault also indicate at least one phase of sinistral movement. The approximately 15° to 20° rotation of strike between the Portwrinkle and Portnadler faults indicates sinistral shearing between them, but in the area between the Portwrinkle and Hoodney Cove [352 539] faults, a 45° clockwise rotation of strike implies a substantial dextral displacement subsequent to the sinistral displacement. Varying cumulative vertical movement components are evident along the faults, with downthrows to north-east or south-west. There is, for example, an estimated 200 m downthrow to the south-west on the Rame Fault between Crafthole [370 542] and Tregantle Fort [385 532].

There are many other mapped dextral strike-slip faults, including the cluster through the Plymouth Limestone Formation, the most prominent of which is the Sutton Harbour Fault, and the Lynher Valley fault zone. The latter, passing through the River Lynher valley to the north of Notter Bridge [384 614], is not a single continuous fault but a north-west-trending zone of faults, locally individually terminated or displaced by cross-cutting faults.

North-east-trending faults in the Landrake and Saltash areas, between Latchbrook and Antony Passage show sinistral movements of several hundred metres by consistent displacements of folded lithostratigraphical units. Associated north-west faults, mapped on the basis of lithology terminations and featuring, are either offset or terminate against the north-east faults.

Strike-slip faults with small displacements are locally very numerous, and north-west-trending dextral faults are observed to occur commonly with conjugate north-east sinistral faults around the Rame Peninsula. Near Penlee Point [4360 4875] a thin dolerite dyke occupying the sinistral dominant member of a set showing minor complementary displacements is cleaved by $S_1$. Continuity of cleavage between dyke and country rock points to formation of the fault set before the end of cleavage development.

A majority of this fault set, oblique to regional strike and including the main named faults, are thought to have developed during $D_1$, but after closure of the basins. Once established they have accommodated various lateral and vertical movements under changing stress regimes. Dextral displacement of the margin of the Bodmin Granite metamorphic aureole by 750 m along the Portwrinkle Fault at Kilham [206 672] indicates post-Early Permian movement, as does the brecciation on the Kingsand Rhyolite along the Cawsand Fault at Kingsand [436 506], but no evidence has been revealed that might further constrain particular movements.

### THIRD PHASE STRUCTURES, $D_3$

Southerly vergent open to close mesoscopic folds of $S_1$, sporadically with an axial plane spaced crenulation cleavage, and occuring as isolated fold pairs or small families of folds, are assigned to $D_3$ (Figure 28, h). A gentle plunge east-north-eastwards is common. The dip of fold axial planes, or axial plane cleavage, differs between localities from gentle to steep to the north-north-west. A more pervasive presence of these folds is evident where $D_2$ structures are well developed to the south of Liskeard. The plot of $S_2$ in the area (Figure 28, g) shows evidence of subsequent rotations, the main one of which is the deformation of the essentially gently inclined $S_2$ into open folds. A steeply inclined crenulation cleavage with appropriate $D_3$ orientation is sparsely developed but a relationship with the folding of $S_2$ has not been determined.

Some of the larger areal reorientations of $S_1$ have a geometry similar to the observed mesoscopic occurrences of $D_3$ folds, but on a much larger scale, for example in the vicinity of Ince and St John, and along the southern extremities of the area in the Rame Peninsula and at Polperro. The southerly verging fold pairs have relatively short common limbs, involving a few hundred metres of strata measured in the dip direction. They dip steeply southwards or are overturned to dip northwards, rotating not only cleavage but also bedding and $F_1$ folds to face southwards. These large structures are interpreted as part of the $D_3$ deformation.

### EXTENSIONAL FAULTING

#### Low-angle faults

Common through the domain are thrusts and low-angle faults exhibiting extension, examples of which are present at Black Ball Cliff [3714 5348] (Figure 29) and an unnamed promontory [437 613] between Neal Point and Parsons Quay (Figure 31). At Black Ball Cliff, a low-angle fault dips gently south-south-east across the 60 m-high face; it separates grey mudstone and sandstone low in the Bovisand Formation in the footwall, from

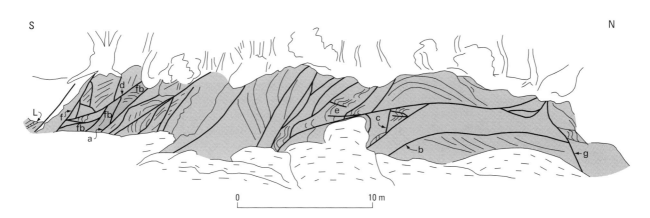

**Figure 31** Fault complex in chert and siliceous slate forming a promontory [437 613] north of Neal Point. Early northerly thrust movements evident on a and b. Steep faults cut or terminated by gently to moderately inclined faults, c and d, and deformed by associated folds indicating southerly down-dip movement (e).

Elsewhere low-angle faults cut by steep normal faults, f and g.  fb fault breccia;  L limestone

stratigraphically higher ochre, red and purple sandstone and siltstone of the Lower Longsands Sandstone Member. Both $F_1$ folded bedding and $S_1$ are steeply inclined in the footwall, whereas bedding is very gently inclined and $S_1$ moderately inclined in the hanging wall. Northward-verging $F_2$ folds in the hanging wall are locally obscured by veining and tension gashes indicating down dip movement. Small-scale structures and relationships indicate the fault to be an out of sequence $D_2$ thrust that transported the sandstone member from the relatively unfolded long limb of a major $F_1$ fold across the steeper rocks of the adjacent antiformal core zone. Extensional back-slip, exceeding thrust movement, then juxtaposed the higher sequence rocks with the lower, such that the fault currently exhibits an extensional offset of bedding. Steep cross-cutting faults with southerly downthrows have accommodated subsequent extension.

At the locality near Neal Point (Figure 31) a fault complex, involving subhorizontal, moderately inclined and steep dislocations, extends over 60 m between the faults bounding the promontory. Small-scale structures associated with the fault breccia on the southern side of the complex indicate early up-dip movements and later down-dip movements. $S_2$, locally an intense shear fabric, is an axial plane cleavage to minor folds cascading northwards in the section. Fault relationships are variable through the zone. Low- and moderate-angle faults locally cut or terminate steep faults, and the folding associated with down-dip movement on the low-angle faults has deformed earlier steep faults. Elsewhere low-angle faults terminate against steeper faults with deformation indicating normal movement on the steeper faults. The main low-angle fault steepens to north and south. It is probable that some of the steeper faults were synthetic faults, associated with the shallower northward climbing thrusts. Subsequent back-slip has locally deformed the steeper structures, and continued extension has produced steep faults cutting the earlier shallower structures.

*Moderate to steep faults*

The fault distribution pattern is dominated by faults trending approximately north-west. A sub-set trending west-north-west is differentiated from the main north-west-trending set, and from the north-north-west set described above (see *Strike-slip faults*). Subordinate sets trend east–west, north–south and east-north-east. The last set (east-north-east) relates locally to the north-east-trending faults described above (p.103) or, where the regional bedding and cleavage strike trends east-north-east, they are steep strike faults.

*West-north-west–east-south-east faults* are closely associated with the north-west-trending faults. In places the major faults (Cawsand and Portnadler faults) have segments of this trend. Prevailing displacements are dextral.

*East–west faults* are common in areas of regional east-west strike. Lying parallel to strike, they are not readily distinguished in areas of poor exposure and few have been mapped. In the area west of Looe, exposed east–west fault planes, oblique to bedding strike, exhibit quartz fibres which indicate several episodes of vertical, oblique and horizontal displacements [e.g. 2111 5078, 2069 5061, 2133 5080].

*North–south faults* generally form inconspicuous features, a notable exception being the fault occupying the East Looe River valley to the north of Looe Mills [231 648]. They carry the late lead-silver mineralisation that was intensively investigated and worked in the district but, without distinctive markers, records do not indicate fault throws. Those faults between the Cawsand and Portwrinkle faults show offsets indicating downthrows to the west. The margin of the Bodmin Granite metamorphic aureole shows an apparent 800 m dextral offset on the north–south fault north of Looe Mills.

*East-north-east–west-south-west faults* appear laterally persistent in areas of lithological contrast, but are not easily distinguished in monotonous lithologies. In the Plymouth urban area, throws on these faults are to the north or south. Pragian sedimentary growth faults of this trend, with a history of dyke intrusion and displacement, are present in the Polperro area [e.g. 2133 5080] west of the Portnadler Fault. Although the general strike of the faults is similar, there is no evidence of a continuous history between such early faults and the late faults of Plymouth.

**Northern Domain**

The domain is defined on the basis of the generally southwards facing of the $D_1$ fold; in the Southern Domain facing is generally northwards. The boundary between domains is not clearly delineated but occurs towards the southern faulted margin of the Tavy Basin sequence just to the north of the Trehills Sandstone Member. At Trehills Plantation Quarry [479 620] the member is mesoscopically folded but the fold envelope is overturned northwards and facing is to the north. Westwards, the dip of the member passes through the vertical to dip to the north and, in the east bank of the River Tavy, to be subhorizontal and upward facing. It thus appears to be part of the steep limb of a major northward-verging structure and part of the Southern Domain. To the north all $D_1$ structures face to the the south although the Trehills Sandstone Member as such is not repeated within the domain. There are possibly equivalent rocks at Hole's Hole [4291 6545] where medium to thin beds of fine-grained sandstone within grey-green silty mudstone define folds facing southwards.

The zone of structural confrontation is located in the vicinity of Cargreen on the west bank of the Tavy. The precise location is uncertain due to poor exposure and to the paucity of marker horizons that could indicate fold facing in the Tavy Formation. Within these constraints, and the presence locally of $D_2$ structures and faults (both steep and low-angle as elsewhere) the zone of confrontation does not display interference between the structures of the Southern Domain and Northern Domain. $S_1$ is common across the zone.

## First major deformation structures, D$_1$

### S$_1$ cleavage

The first phase cleavage is slaty in the fine-grained lithologies that dominate the Tavy, Torpoint and Burraton formations, and is a well-developed penetrative fabric in the rarer coarser grained rocks. It shows a considerable dip variation largely reflecting later deformation, and variation in strike is less pronounced than in the Southern Domain. The attitude and orientation of S$_1$ is essentially bimodal. The principal mode indicates a gentle to moderate dip southwards, and the subordinate mode indicates gentle to moderate dips northwards. Near-vertical cleavage with dip azimuths between north-west and north-north-east and between south-east and south-south-west is also present; a prominent zone extends between Have [427 634] and Ellbridge [402 630].

### F$_1$ folds

Few F$_1$ folds are recorded in the Northern Domain, principally because of the lack of lithogical differentiation in the Tavy Basin. The sporadically developed arenaceous beds indicate either minor folds, or they are present as small hinge zones or parts of limbs isolated by dislocation along cleavage. A notable exception is an occurrence at Hole's Hole [4291 6545] where a packet of thin to medium beds of fine-grained sandstone with greenish grey, silty mudstone interbeds define close folds about slaty cleavage inclined moderately southwards (Plate 18). The sequence is essentially inverted and the folds face southwards and downwards. Southerly facing is indicated elsewhere through the domain by grading and cross-lamination cut-off in minor folds and bedding/cleavage intersection lineations.

## Second major phase deformation structures, D$_2$

S$_2$ cleavage is a crenulation cleavage that dips to the south-south-east at moderate to steep angles. It is axial planar to northward verging, asymmetric, open to close folds of S$_1$ that plunge gently between east-north-east and east-west. F$_2$ folds have gently to moderately inclined, southerly dipping, long limbs; the short limbs are vertical, dip steeply southwards, or have a steep to moderate northerly dip. Structures are on both mesoscopic and macroscopic scales, and much of the near-vertical and southerly inclined S$_1$, may be attributable to D$_2$.

## Third phase structures, D$_3$

Folds and cleavage of D$_3$ age are, like those of D$_2$, developed only sporadically. Their temporal relationship with earlier structures is evident in the section on the east bank of the River Tavy between Warleigh Quay [4553 6168] and the Lime Kiln at Sandgore Lane [4599 6228]. There folds are predominantly minor, asymmetrical buckles plunging gently south-eastwards and verging south-westwards with a spaced axial plane crenulation cleavage (Figure 28, h). Cleavage is also present over larger areas without associated small-scale folds, in places intersecting D$_2$ structures. Some 360 m north-east of Warleigh Quay in mudstone with silty laminae, a large mesoscopic, northward verging, close, F$_1$ fold couplet is folded by a southward-verging, open, F$_3$ couplet of similar scale and transected by the associated crenulation cleavage that is gently inclined northwards. S$_1$, axial plane cleavage to the F$_1$ folds, is reorientated progressively from a moderate southerly dip through the vertical to a northerly dip and then back to the south over several tens of metres of the river cliff section through the F$_3$ folds. Small-scale asymmetric F$_3$ folds cascading to the south are developed in the lower long limb of the F$_1$ fold pair, the synclinal hinge of that pair is opened out, and the formerly overturned common limb of that structure is refolded to be predominantly the right way-up.

## Faulting

Faulting with a number of trends has affected both the rocks of the Northern and the Culm domains. Apparent displacements are influenced by the flat-lying nature of the lithostratigraphical units within the tectonic slices of

**Plate 18**  Closely folded sandstone beds within the Tavy Formation at Hole's Hole [4291 6545]. The folds face southwards and downwards. Looking eastwards (GS526).

**Table 15**  Summary of fault displacements in the St Mellion outlier.

| Trend | Displacement | Relative age |
|-------|-------------|--------------|
| NW–SE | downthrow to NE and SW | truncates thrusts and E–W faults truncated by NE–SW faults |
| NE–SW | downthrow to NW | truncates thrusts and NW–SE faults truncated by ENE–WSW faults |
| WNW–ESE | downthrow to SSW | truncates thrusts |
| ENE–WSW | downthrow to NNW | truncates thrusts and NE–SW faults |
| E–W | downthrow to S | truncates thrusts truncated by NE–SE faults |

the St Mellion outlier which provide good markers for vertical displacements but only poor control on lateral displacements. Within this constraint fault trends, chronology and displacements are listed in Table 15.

## Culm Domain

This domain comprises rocks of the mainly Dinantian St Mellion, Brendon and Newton Chert formations. Although structures within these formations are similar to those in the subjacent Devonian formations of the Northern Domain, the Culm Domain is distinguished on the bases of its evident allochthonous and dismembered characteristics (see p.69) and generally lower intensity of first-phase tectonic fabric development.

### First phase structures, $D_1$

#### $F_1$ folds

Mesoscopic and macroscopic folds are present within the flysch facies rocks of the St Mellion and Brendon formations. Small-scale gently inclined to recumbent folds are close and asymmetric, with short limbs down to 0.5 m and long limbs in excess of 2 m. Facing is to the south and vergence is variable between north and south. Large mesoscopic folds are observed where the scale of exposure is sufficient, as in the Paynter's Cross road cutting [397 645] (Figure 21). The structure at that locality indicates that those folds are parasitic to major folding, and the presence within the St Mellion Formation of overturned folds with kilometre-scale wavelengths can be inferred from the distribution of right way-up and inverted graded beds.

#### $S_1$ cleavage

The first cleavage of the domain is well developed within the hinge zones of $F_1$ folds where it is penetrative in both sandstone beds and interbedded siltstone and mudstone. Cleavage intensity diminishes away from hinge zones, preferentially within the sandstone beds, such that in places cleavage is not evident in either the sandstone or finer rocks. The bedded chert rarely displays any cleavage fabric, but an incipient recrystallisation produced sericite showing a bedding-parallel preferred orientation (E 67270). $S_1$ shows considerable

variation in strike and dip about a modal gentle inclination southwards.

### Post $D_1$ folding and cleavages

Post $D_1$ folds and cleavages are sporadically developed within the domain but structures of differing orientation and style have not been observed to interfere. Their phase numbers are based on comparison with structural style elsewhere.

#### $D_2$ structures

Second phase structures comprise locally developed, minor, open to close, buckle folds of bedding and $S_1$ cleavage, with a variably developed axial plane cleavage that is a crenulation cleavage in fine-grained rocks and a fracture cleavage in sandstone. The folds trend about the east-north-east azimuth, and are asymmetrical with shorter northern limbs. $S_2$ cleavage is gently to moderately inclined southwards, but the attitude of limbs is dependent upon that of previously folded bedding and $S_1$ (see Figure 21).

#### $D_4$ (?) structures

$S_1$ is very locally deformed into small-scale, east-north-east-vergent, asymmetric folds with, in places, an axial plane crenulation cleavage dipping moderately west-south-westwards. The folds are open to close with short limb lengths of less than 1 m and plunge gently south-south-eastwards. These structures are restricted to an area along the west bank of the Tamar, north of Halton Quay [4135 6548]. Their relative age within the structural sequence is speculative but their geometry displays strong similarities to folds and cleavage assigned to $D_4$ on the north Cornish coast by Goode and Leveridge (1991).

### Thrusting

Boundaries to the lithostratigraphical divisions of the St Mellion outlier are generally flat lying to gently inclined; they can be constrained closely, but are not normally exposed. In a temporary exposure on Vinegar Hill [3980 6404] the junction between subhorizontal chert beds of the Newton Chert Formation above and near-vertically cleaved mudstone of the Tavy Formation below is 1.0 to 1.5 m of yellow clay fault-gouge, dipping southwards at 12°. Such contrasts in attitude of bedding and/or

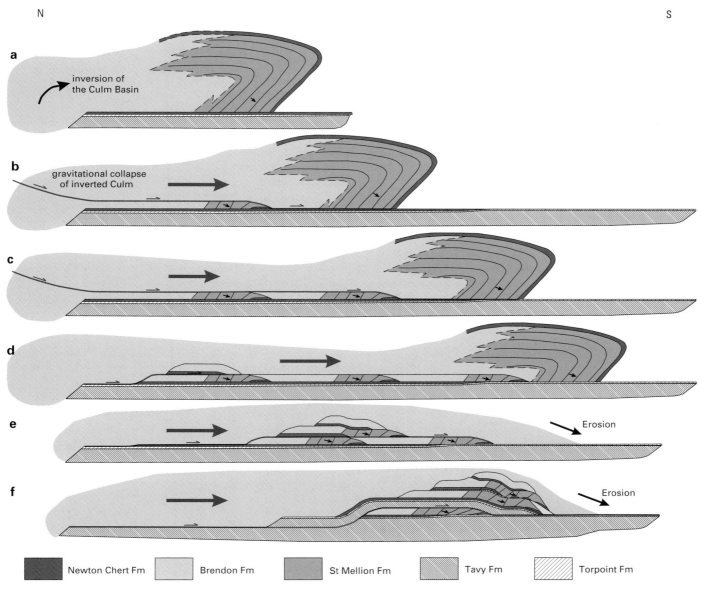

N

S

a inversion of the Culm Basin

b gravitational collapse of inverted Culm

c

d

e Erosion

f Erosion

| | Newton Chert Fm | | Brendon Fm | | St Mellion Fm | | Tavy Fm | | Torpoint Fm |

**Figure 32** A tectonic model for the deformation and emplacement of the St Mellion outlier.

a. Inversion of the Culm Basin took place over a prolonged period of time, commencing in the south of the basin in Namurian times. The initial stages of this inversion caused substantial uplift of the sedimentary fill of the basin at its southern edge. The expulsion of the fill from the basin, accompanied by gravitational collapse of the uplifted mass initiated an asymmetric south-facing regional fold in the Culm sediments.

b, c, d. The continued inversion and gravitational collapse caused the Culm sediments to be transported southwards over the sediments of the Tavy Basin on a low angle extensional fault. Failure near the base of this transported mass caused fragments of the upside-down limb of the fold to be isolated as a series of 'horses' along the basal extensional fault.

d, e. Continued inversion and gravitational collapse of the more central and northern parts of the Culm Basin during the $D_2$ deformation in Westphalian times increased the loading on the rear-most parts of the basal extensional fault causing a new failure plane to develop within the cherts at the top of Tavy Basin fill. Strike-slip movements of this new failure plane caused it to connect to the still-moving parts of the original extensional fault via a series of ramps. As movement occurred along the basal movement plane the isolated Culm 'horses' were stacked sequentially, forming a duplex structure.

f. Continued inversion and collapse of the uplifted Culm sediments caused failure within the sediments of the Tavy Basin. This new failure plane transported the duplex, together with a slice of the Tavy Formation, over one of the isolated Culm 'horses' forming the stacked sequence seen in the St Mellion outlier.

cleavage across most boundaries indicate that they are most probably faulted. The prevailing inverted bedding of the St Mellion Formation, in common with most of the Silesian rocks of the Tavistock district to the north (BGS unpublished data), is inconsistent with simple southward overfolding of the Culm. An explanation compatible with structural relationships observed is presented in Figure 32 which invokes post $D_2$ gravitational collapse of the inverted Culm Basin deposits, accommodated on flat-lying thrusts, during the late Carboniferous.

## Domain structure correlation

It is evident that the first and second phase structures of all domains compare on the bases of fold style and orientation. They differ in terms of their structural facing, which in the Southern Domain is to the north and in the Northern and Culm domains is to the south. In the zone of structural confrontation there is no evidence of interference, or of an age difference between the first phase structures to north and south. North of the confrontation, the domains differ in that the metamorphic grade associated with cleavage formation is lower in the Culm Domain than in the subjacent Northern Domain. $D_3$ structures are common to the Southern and Northern domains. $D_4$ (?) is recognised only in the Culm Domain. The significance of the correlateable sequences is that the early main deformation is almost certainly the same deformation in all domains, with structures having a common geometry and a non-interfering relationship.

## Deformation kinematics

The predominant structures of the area are those of $D_1$. In the Southern Domain, $F_1$ folds to the east of the Portwrinkle/Rame fault have an approximate east-west trend, and those to the west have a spread about the east-north-east azimuth. Fold asymmetry is consistent with overriding transport to the north and north-north-west, respectively. The latter is parallel, in areas of incipient sheath folding, to the bisectrix of fold spread and recorded extension lineation (Figure 28, d). The transport represented by $F_1$ folds in the Northern and Culm domains is directly opposed, being to the south.

Second phase folds show a wide trend variation which does not appear to be a function of sheath folding, with pitch angles being small irrespective of trend, but due to variable initial orientation dictated partly by earlier bedding and $S_1$ fabric attitudes. Fold trend spread about east-north-east and west-south-west azimuths appears to be similar on both sides of the Portwrinkle/Rame fault suggesting that reorientation of $D_1$ structures across that line developed before or during $D_2$. $F_2$ fold asymmetry indicates overriding movement to the north-north-west. Strain coordinates were thus close to those of $D_1$ structures between the Portwrinkle and Portnadler faults and slightly oblique elsewhere.

$D_3$ and $D_4$ folds of $S_1$ are modified buckle folds with asymmetries indicative of transport southwards and to the east-north-east, respectively

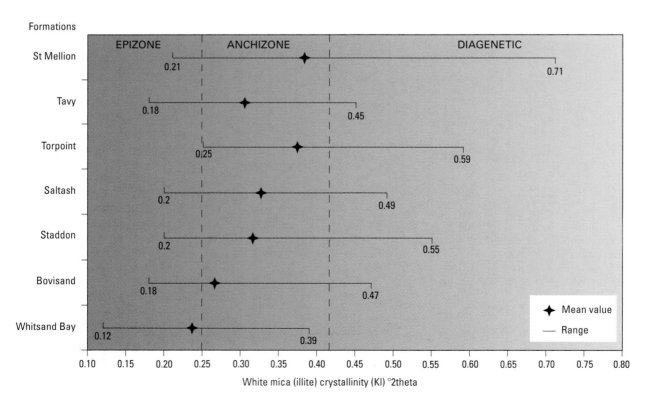

**Figure 33**  White mica crystallinity values for the district.

# METAMORPHISM

The rocks formed within the Variscan sedimentary basins of south-west England currently exhibit low-grade metamorphism between late diagenetic and epizone grades (Warr et al., 1991 and references therein). Within the Plymouth district, the metamorphic grade of pelitic rocks has been determined from illite crystallinity measurements (Figure 33) and these exhibit a general trend of decreasing metamorphic grade with stratigraphical younging. The data from the Tavy Formation, however, show anomalously high metamorphic grades, for its stratigraphical position. This probably reflects thermal overprinting within the metamorphic aureole of the Dartmoor Granite (see below) which affects a high proportion of the Tavy Formation in the north-east of the district.

## Regional metamorphism

The general trend of grade increasing with increasing age implies that a pattern of burial metamorphism may have been acquired at an early stage in the tectonic development of the district. This early depth-related pattern was progressively modified by deformation and locally overprinted by contact metamorphism. Post-metamorphic faulting has disrupted the regional metamorphic pattern, offsetting it along the major faults and causing smaller irregularities due to block faulting (Figure 34). To the west of the Portwrinkle Fault, the pattern of metamorphism generally shows grade increasing towards the south-east into the oldest strata, typically high anchizone in grade with the epizonal grade rocks exposed in the cores of regional folds. In the northern part of this area, however, the regional pattern is locally overprinted by the thermal metamorphism associated with the Bodmin Moor Granite. East of the Portwrinkle Fault, the regional metamorphic pattern has been offset dextrally by strike-slip movements on the fault, resulting in a widened crop of low anchizonal grade rocks and displacement of high anchizone–epizone grade rocks towards the south-east. Late diagenetic rocks around St Mellion are representatives of the youngest strata in the district which were tectonically emplaced southwards over higher grade rocks during the post-metamorphic inversion of the Culm Basin (see chapter 11). The overprinting of these diagenetic rocks, as well as the underlying anchizone grade rocks, by contact metamorphism of the Dartmoor and Bodmin granites indicates that inversion of the Culm basin pre-dated granitic intrusion.

All the metabasic volcanic rocks in the district show some degree of alteration, but only 20 per cent contain

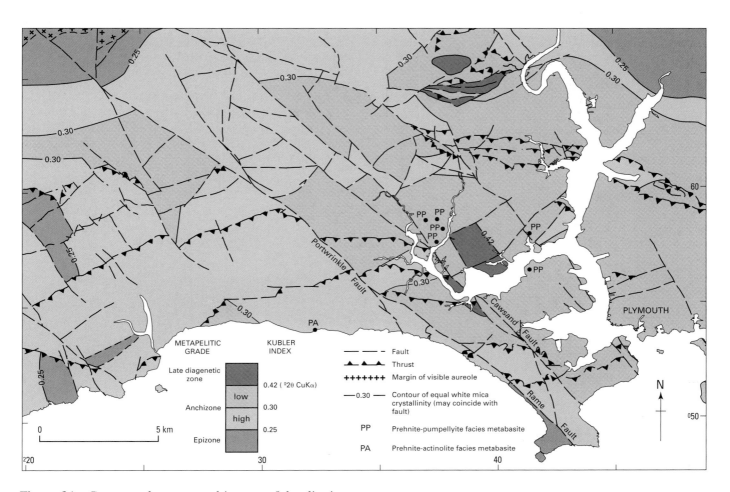

**Figure 34** Contoured metamorphic map of the district.

minerals which are indicative of a very low-grade metamorphic facies (Merriman et al., 1996). These diagnostic minerals are indicative of prehnite–pumpellyite facies in two areas in the district (Figure 34). South of Saltash, they occur at Antony Park [415 567], Antony Passage [413 579] and around Forder Lake [414 579] and farther to the west around St Germans [375 577] and St Erney [360 589]. In terms of metapelite zones, both areas are within the low anchizone, consistent with the widely accepted correlation between the prehnite-pumpellyite facies and the anchizone (Kisch, 1987). A dolerite sill containing a prehnite-actinolite facies assemblage crops out in cliffs at Long Stone, about 1.6 km west of Portwrinkle, intruding metapelitic rocks of the high anchizone (Figure 34).

The pattern of regional metamorphism developed in the Plymouth district suggests that very low-grade (burial) metamorphism was initiated under an overburden tectonically thickened to the south by over-thrusting from the south as the Gramscatho Basin closed in the latter part of the Late Devonian–Early Carboniferous (Leveridge et al., 1990). Progressive inversion of the passive margin basins (see chapter 11) preserved the lowest grade rocks in the frontal region of the advancing thrust wedge. Here the pattern of very low-grade metamorphism is characterised by late diagenetic to low anchizonal grades in mudrocks and prehnite-pumpellyite facies in basic volcanic rocks. These grades imply minimum temperatures of 175°C for the prehnite-pumpellyite grade in metabasic rocks (Frey et al., 1991), and approximately 200°C for the late diagenetic–anchizone boundary (Kisch, 1987). Under a normal geothermal gradient of 25°C/km these metamorphic conditions require at least 7 to 8 km of overburden. However, on the basis of potassium mica b lattice dimension studies, Warr et al. (1991) suggest that the regional metamorphism in south Devon was of a low-pressure facies type and inferred a geothermal gradient of 40°C/km. This suggests an initial overburden thickness of 4 to 5 km.

As overthrusting of the rocks of the district proceeded from the south, frontally accreted very low-grade rocks were progressively buried and deformed, and incorporated in the thickened wedge. During this phase of tectonism, fold tightening and the development of a penetrative slaty cleavage would have promoted strain-enhanced crystal growth of white micas (Merriman et al., 1995), resulting in high anchizonal to epizonal metapelites. Prehnite-actinolite facies metabasic rocks associated with these metapelites suggest that peak metamorphic temperatures did not exceed 320°C. On the basis of a 40°C/km geothermal gradient such temperatures indicate that up to 8 km of crustal thickening resulted from thrust stacking. This may be a minimum thickness since Warr et al. (1991) suggest that a change to lower geothermal gradients is associated with thrust stacking in the Plymouth area.

## Contact metamorphism

The outer aureoles of the Bodmin and Dartmoor granites, delineated by the KI = 0.30 isocryst (see Figure 34), extend well beyond visible hornfelsing into the northern part of the district. Areas of epizonal rocks in the north-west and north-east corners of the district are parts of the aureoles of the Bodmin Moor and Dartmoor granites respectively. Two areas of high anchizonal grade rocks, one extending south-eastwards between the Portwrinkle and Cawsand faults through Menheniot [289 629] to Tilland [320 618], and the other along the Lynher Fault zone to Notter Bridge [385 608], were possibly caused by fault controlled channelling of fluids from the Bodmin Granite (see chapter 3). The mapped aureole in the north-west corner of the sheet shows dextral displacement along the Portwrinkle Fault and dextral displacement or westerly downthrow on north–south faulting.

ELEVEN

# Tectonic geology

The rocks of the district lie within the Variscan orogenic belt (see Franke, 1989). In the British Isles the orogenic belt affects rocks in southern England and small areas of southern Wales and Ireland, but within Europe the deformed rocks are very extensive, forming the sub-Mesozoic basement of France, Spain, Portugal, Germany, Switzerland and parts of Austria, Poland and the Czech Republic. The Plymouth district lies in the outer part of this broader area, within the Rhenohercynian Zone (Kosmat, 1927), and contiguous with the rocks of this zone in Belgium and Germany (Holder and Leveridge, 1986a).

The plate tectonic implications of Kosmat's (1927) long standing division of the Variscan rocks in Germany, Poland and the Czech Republic, into the Rhenohercynian, Saxothuringian and Moldanubian zones were not realised until recently. These zones were thought to represent intracontinental fold belts (see Martin and Eder, 1984; Matthews, 1984) with the various facies of the Devonian–Carboniferous succession assigned to basin or rise environments within a zone of essentially vertical tectonics (Schmidt, 1926), or interpreted as lying within a zone of transpression with localised basins (Badham, 1982). Even the presence of an oceanic suture in the Massif Central, proposed by Bard et al. (1980), has been confirmed only in recent years by the identification of ophiolitic complexes within the major nappes (Dubuisson et al., 1989). Recent reinterpretation of the Rhenohercynian, Saxothuringian and Moldanubian zones in Germany, Poland and the Czech Republic and their interpretation as oceanic collision belts (see Franke, 1989), together with the linking of the Variscan Massifs in Europe has led to a reinterpretation of the European Variscides in terms of a classical collision model (see Holder and Leveridge, 1994).

The sedimentary rocks of the Plymouth district and their correlatives in south-west England were deposited in fault-bounded basins which are recognised as forming part of a passive margin sequence (Holder and Leveridge, 1994). The correlation of these rocks across north-western Europe (Holder and Leveridge, 1986a) suggests a pattern of extensive east–west-trending sedimentary basins forming the northern passive margin of a laterally extensive Rhenohercynian oceanic basin (Franke, 1989). Oceanic remnants and active margin sequences from that basin are preserved in southern Cornwall. The sequence of extension and basin opening followed by inversion and deformation within the Variscan rocks of the Plymouth district, recorded in this memoir, is related to the continental rifting and growth of the Rhenohercynian Ocean followed by its closure during continental collision. It thus represents the key elements of a classic 'Wilsonian orogenic cycle' (Wilson, 1966;

Smith, 1993). Rifting of the passive margin occurred over a period of approximately 65 Ma with extensional effects moving sequentially northwards affecting a progressively larger zone. The lack of evidence for a later sequence of sediments mantling the rift basins in the district may indicate that this was a starved passive margin or that thermal subsidence associated with the post-rifting stages of ocean spreading (Bott, 1992) was not completed before continental collision occurred. The contemporaneity of inversion in the Gramscatho Basin (Holder and Leveridge, 1986b) and the subsidence of the Culm Basin, lends support to the latter hypothesis.

Deformation of the Rhenohercynian passive margin rocks was related to the latest of several continental collision events within the Variscan deformation belt of southern and central Europe (see Holder and Leveridge, 1994). In the Plymouth district this collision-related deformation progressed northwards with time by a process of inversion, out-thrusting of basin fill on basin margin faults, and deformation until the faults locked and stress could be transmitted to the next basin. This style of 'soft' deformation compartmentalised compressive stresses, allowing extension and sedimentation to proceed in the north of the district as basins to the south were being deformed. The collision-related deformation was therefore a relatively slow and continuous process which spanned approximately 50 Ma, becoming diachronously younger northwards.

The geology of the Plymouth district, the stratigraphy of its sedimentary rocks, its igneous rocks, mineralisation, structure and metamorphism can be integrated to elucidate the development of the Variscan passive margin and its subsequent evolution.

## BASIN DEVELOPMENT AND SEDIMENTATION

The Rhenohercynian passive margin sequence within the Plymouth district was deposited in four major east–west-trending fault-bounded basins; the Looe Basin, South Devon Basin, Tavy Basin and Culm Basin and on intervening highs. Although there is a degree of contemporaneity between adjacent basins the age of the sedimentary rocks infilling the basins shows a general decrease from Early Devonian in the southernmost Looe Basin to the Carboniferous in the Culm Basin, the northernmost represented in the district.

### Looe Basin

The Looe Basin is the most southerly basin of the district and probably of the passive margin, being overthrust to the south by the flysch of the Gramscatho Basin (Holder and Leveridge, 1986b). Its northern boundary is defined by the inter-basinal high deposits of the Plymouth Limestone Formation. It includes the Jennycliff division of the Saltash Formation, but overthrusting of the Staddon Formation has obscured this part of the basin west of the Cawsand Fault. Along much of the east–west strike of the basin in Cornwall, therefore, the Looe Basin is represented solely by the rocks of the Dartmouth and Meadfoot groups.

The strata of the Looe Basin constitute three major thrust nappes. At each thrust front, large-scale antiformal folding is present in the hanging wall, which suggests that internally the Looe Basin (Figure 35) comprised three sub-basins with half-graben geometries (see p.99 and Figure 39). These are termed, from south to north, the Freathy, Long Sands and Duloe sub-basins. The apparent decrease in the proportion of the older Dartmouth Group rocks from the most southerly to the most northerly of these sub-basins, together with the appearance of the youngest rocks only in the most northerly of the sub-basins suggests that they were not

formed synchronously but developed from south to north.

The formation of the Looe Basin appears to have begun in Pragian or possibly Lochkovian times. The earliest rocks representing the basin fill, the Whitsand Bay Formation, lie within the Freathy and Long Sands sub-basins. The Whitsand Bay Formation rocks were deposited in fresh water lacustrine and fluvial environments in basins in which subsidence was episodic and controlled by extensional fault movements (chapter 4; Smith and Humphreys, 1989, 1991). Episodic down-faulting was accompanied by the deposition of quartzitic sheet sandstone beds (Jones, 1993), occasionally accompanied by an incursion of grey marine mudstone, followed by the resumption of lacustrine purple mud deposition and then grey-green fluvial and lacustrine mud and sand deposition.

Rocks of the Bin Down Formation within the Long Sands sub-basin are of similar age to the upper part of the Whitsand Bay Formation with which it interdigitates. They were deposited in a similarly tectonically active

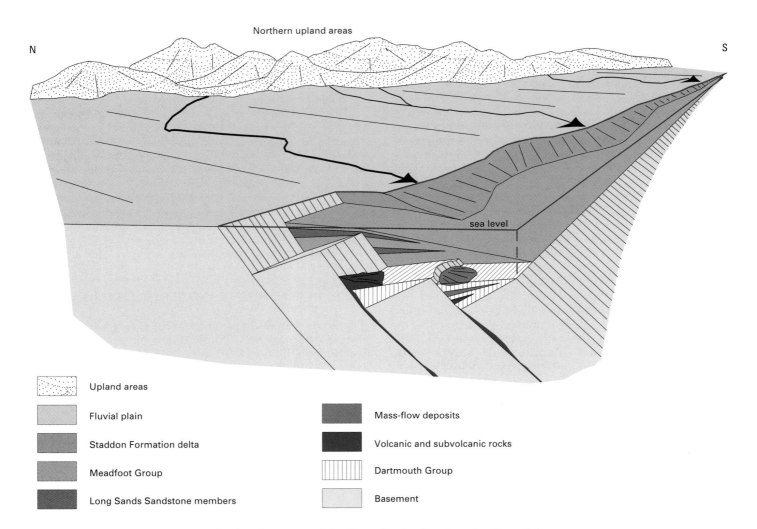

| | |
|---|---|
| Upland areas | |
| Fluvial plain | Mass-flow deposits |
| Staddon Formation delta | Volcanic and subvolcanic rocks |
| Meadfoot Group | Dartmouth Group |
| Long Sands Sandstone members | Basement |

**Figure 35** A synoptic view of the development of the Looe Basin showing the disposition of the main lithostratigraphical divisions.

environment. The Bin Down Formation, apparently representing a major marine ingress, is, however, of a more sandstone-dominated facies and includes a substantial thickness of subaqueous volcanic rocks. These features suggest its geographical separation from the Dartmouth Group rocks of the Freathy sub-basin.

Contemporaneous volcanic rocks within the sub-basins are alkaline and calc-alkaline basalts (see p.89). Trace and rare-earth-element geochemistry and negative $\Sigma_{Nd}$ (T) values of the calc-alkaline basalts indicate crustal contamination of mantle-derived magmas, interpreted to represent the early stages of intracontinental rifting (Merriman et al., 2000).

The onset of full marine conditions in the later Pragian with the deposition of the Bovisand Formation indicates deepening of the Looe Basin as subsidence exceeded the rate of sediment deposition. The continued extension causing this subsidence generated the Duloe sub-basin as a new half-graben to the north of the Long Sands sub-basin. Between the Pragian and the Middle Emsian, environments in both the Long Sands and Duloe sub-basins were similar. Two major polycyclic sequences, each cycle comprising offshore shelf mudstone and siltstone succeeded by sandstone deposited in a wave or storm dominated shoreline environment, represent episodes of subsidence and basin filling. Contemporaneous rocks of the Meadfoot Group within the Freathy sub-basin, include turbidites that may have been deposited in a slightly deeper-water setting.

In the Duloe sub-basin, by Emsian times, the deposition of the delta-top to fluvial sandstone beds of the Staddon Formation indicates that at the northern edge of the Looe Basin the rate of sediment deposition had outstripped subsidence with rivers feeding clastic sediments into the basin across a piedmont from a continental area to the north.

Regional extension, renewed towards the end of the Emsian, produced more subsidence within the Looe Basin and initiated the South Devon Basin to the north. This isolated a 'high' between the basins, which shielded the southern basin and prevented coarse clastic sediments from being deposited there. The consequent change in the Looe Basin environment is reflected in the Jennycliff division of the Saltash Formation. Sandstone and siltstone of Emsian age in its lower part indicate a continuing but restricted clastic sediment supply, removed during storm activity from the drowned coastal fluvial plain, as basin subsidence resumed at the end of the Staddon Formation deposition. Thereafter, during the early Eifelian there was a rapid decrease in the proportion of sandstone and the appearance, in the mudstone-dominated sequence, of beds of bio-clastic limestone. Sedimentation of a starved basin facies continued in the Looe Basin to the top of the Jennycliff division in the late Eifelian, with an increasing proportion of limestone storm deposits derived from the carbonate banks formed in the shallows above the developing Plymouth High. Basic volcanic rocks in the upper part of the sequence became a significant component of marginal deposition.

The geochemistry of the volcanic rocks within the Looe Basin reflects an active rifting environment. Alkaline and limited calc-alkaline basaltic hyaloclastite,

tuff and lava are most common in the earlier deposits of the basin, particularly within the Bin Down Formation, and are relatively uncommon within the succeeding rocks of the Bovisand Formation. According to White (1992) the lack of significant volumes of volcanic rocks in a rift environment reflects a slow rate of crustal extension which is typical of the early stages of continental rifting. Nevertheless, the thick deposits of the basin, for example over 3300 m of Pragian perennial lake deposits of the Whitsand Bay Formation, do indicate active subsidence and sediment supply.

## Plymouth High

The South Devon Basin formed by rotational faulting on its northern boundary during the Eifelian. It was a half-graben basin and the southern edge of the rotated fault block formed a linear structural high separating the South Devon Basin from the Looe Basin to the south. This feature, the Plymouth High, is now represented in two relatively thin thrust slices that make up the Plymouth Limestone Formation. To the east of the district, limestone associated with this inter-basin high extends across south Devon to Torbay. To the west, however, increased northward displacement on the out-of sequence thrust bounding the Staddon Formation has obscured the Plymouth Limestone Formation beneath the overthrust sedimentary fill of the Looe Basin.

Deposition on the Plymouth High evolved from scattered crinoid colonisation of a mud-dominated bank in the early Eifelian, through a carbonate bank in the mid to late Eifelian, to a stromatoporoid and coral reef and patch reef complex in the Givetian and Frasnian. The extensional activity which formed the South Devon Basin and caused renewed subsidence in the Looe Basin is also recorded on the Plymouth High by extrusive volcanic rocks. Continuing activity along bounding faults is reflected in the thick local volcanic expressions in the late Eifelian through the Givetian and into the Frasnian. Water depths over the high appear to have been within storm wave base throughout the Givetian. In early Frasnian times, water depths became shallower; the reef migrated southwards and back-reef limestone was deposited over the older parts of the reef itself. This migration appears to have been caused by the rate of reef growth exceeding the rate of subsidence of the Plymouth High during the early Frasnian. By late-Frasnian times reef growth on the Plymouth High had been abandoned and the reef began to fracture and founder. The abandonment of the reef appears to have been caused by its inundation by clastic sediments overflowing from the South Devon Basin. Red mudstones of the Torpoint Formation appear as thin partings within the Plymouth Limestone Formation in the late Givetian and as Frasnian to early Famennian fillings within fissures in the top of the reef itself. The overtopping of the Plymouth High is also evident from the presence of interbedded purple and green slaty mudstones of Frasnian age (the Falmouth Series of Hill and MacAlister 1906) amongst the flysch facies rocks of the Portscatho Formation in the Gramscatho Basin (see Leveridge et al.,

1990). Selwood et al. (1984) proposed the same mechanism to explain the abandonment of the Ogwell and Chercombe Bridge limestone units, possible lateral equivalents of the Plymouth Limestone Formation to the east in the Newton Abbot district. Those limestone units were inundated in Famennian times by purple and green mudstone (the Whiteway Formation of Selwood et al., 1984) essentially similar to the Torpoint Formation, suggesting that the overflowing of sediments from the South Devon Basin was not a local event, but was related to the filling of the basin at this time.

## South Devon Basin

The South Devon Basin was described by Selwood (1990) as being constrained to the south during the Devonian by platform carbonates on an 'outer shelf', and to the north, at the edge of an 'inner shelf', by volcanics associated with the marginal fault of a half-graben. This broadly compares with the Mid and Late Devonian basin north of Plymouth and the terminology is adopted here. A contiguity with the Trevone Basin (Matthews, 1977) of North Cornwall, similarly modelled by Selwood (1990) but without the carbonate rise, is evident in the Plymouth district.

An immediate consequence of northward fault propagation and initiation of the South Devon Basin (Figure 36) in late Early to early Mid Devonian, was the starvation of the Looe Basin of coarser Staddon-type clastic sediments. The presence or absence of such sediments in the new basin is not established, but a general paucity of coarse clastics in younger Devonian sequences of the district indicates

regionally sigificant, tectonically controlled changes between source and sedimentation regimes at the time of formation of the South Devon Basin.

During the early stages of formation of the South Devon Basin, in the Eifelian and early Givetian, sandstone turbidites with sublittoral shelly faunas, limestone turbidites from biogenic banks and limestone sedimentary breccias off the nascent Plymouth reef were deposited in the south. In the extensional regime of general subsidence, and the added rotational subsidence, there was continuity of the marine sedimentary regime across the Looe and South Devon basins. This is confirmed by the persistent background grey silty mudstone lithofacies of the Saltash Formation, and the nature and age of the early coarse clastic inputs (of the St Keyne and Insworke divisions) into the the South Devon Basin which are similar to those of the Jennycliff division south of the Plymouth High. The segregation of Saltash and Torpoint facies is interpreted as a function of basinal subdivision (see p.65; Figure 36) but it appears that the northern margin of the basin was faulted at an early stage. This is certainly the case if the Givetian conodonts in limestone turbidites in the Landulph division of the Saltash Formation do represent remobilisation of penecontemporaneous carbonate deposits, as their preservation suggests, and not reworking in the Late Devonian. That is to say, the sourcing Landulph High was either developing or part of a relatively elevated shelf area in the Givetian. Volcanicity was relatively sparse at the northern margin of the basin throughout its history, but common in the Eifelian and

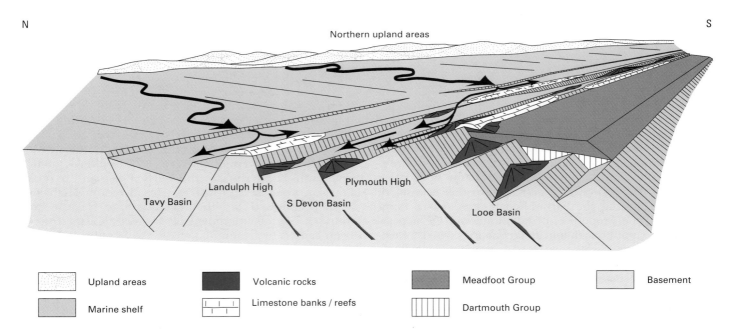

**Figure 36**   A synoptic view of the development of the South Devon and Tavy basins, showing the disposition of some key lithostratigraphical units, and indicating provenance and submarine transport of Torpoint facies sediments.

Givetian to the south on and about the Plymouth High. However the main activity was in the central part of the basin, largely in the Saltash division of the Saltash Formation, from the late Givetian through the Frasnian and into the Famennian. There during this period, and despite disturbance by high-level intrusive and extrusive volcanicity and evident persistence of local centres producing a submarine topography, the finest grained sediments of the formation predominate.

The northerly migration of volcanic activity is interpreted as being concomitant with the development of intrabasinal rifting, and its persistence as representing continuing extension on the intrabasinal faulting. The consequent formation of northern and southern sub-basins is reflected through the Upper Devonian by the segregation of Torpoint Formation rocks to north and south by the Saltash division which represents deposition on a relative high. The presence within the Saltash division of the Wearde Sandstone, deposited during the late Frasnian when the sub-basins were segregated, may possibly be explained by its extrabasinal provenance, and its form as interconnecting channel fill deposits.

The distribution of the Torpoint Formation, and the Plymouth division in particular, interdigitating with the Saltash Formation and petering out westwards, indicates westerly transport in the sub-basins of the continental-derived ferric, iron-rich sediment. It is uncertain whether the inferred westerly palaeoslope within the basin persisted to the shielded deep-water of the Trevone Basin of North Cornwall (R W O'B Knox in Goode and Leveridge, 1991). Selwood (1990) proposed the presence of the Liskeard High between the two basins largely on the basis of interpretation by Burton and Tanner (1986) of their Rosenun Slate Formation as shallow shelf deposits. These Eifelian to mid-Givetian rocks are here regarded as early marginal deposits of the South Devon Basin.

The presence of the Landulph High (Figure 36), that separated the South Devon Basin from the Tavy Basin to the north, is recognised from the Middle and Upper Devonian detrital limestone and nodular cephalopodic limestone in the thrust slices that make up the Landulph division of the Saltash Formation. Derived crinoids, shelly fauna and carbonate mud provide evidence of a carbonate platform, although not reef development, that persisted well into the Famennian, that is after the inundation of the Plymouth High. Instability about the rise is also indicated by slumping and derived microfaunas in the mudstones. The thrust slices, cut from the slopes of the high, are northerly vergent, and the overthrust high, now occluded, was formerly to the south of the Landulph division. The presence of probable Tournaisian chert and possible Viséan chert in sequence within the same thrust stack indicates their former extent across the high and probably the adjacent basins.

## Tavy Basin

The formation of the Tavy Basin to the north of the South Devon Basin (Figure 36), by continued regional extension, appears to have commenced in the Frasnian and to have continued into the earliest Tournaisian.

Sedimentation within it was in large part synchronous with that in the South Devon Basin, although the different lithostratigraphical sequence within these two basins suggests that they were, for the most part, separately sourced. The southern margin of the Tavy Basin is now concealed beneath thrust sheets derived from the northern edge of the South Devon Basin, but it originally lay to the north of the Landulph High. Much of the Tavy Basin is overlain by southerly directed thrust sheets originating on the southern margin of the Culm Basin. The northern margin of the Tavy Basin is most probably indicated by the presence of nodular cephalopod limestones, derived from a local high, within the Whitelady Formation (Isaac, 1983) of Lydford Gorge [502 835], a lateral equivalent of the Tavy Formation some 17 km north of the district. This gives a pre-Variscan deformation width to the Tavy Basin of at least 22 km. The distribution of Tavy Formation facies as an east–west-trending belt across south-west England (as the Kate Brook Slate of Selwood et al., 1984, and the Tredorn Slate Formation of Selwood et al., 1998) suggests that the Tavy Basin trended east–west and that it was probably controlled by east–west trending extensional faults similar to those responsible for the extension of the South Devon and Looe basins to the south. The southerly vergence of structures in most of the rocks of the Tavy Basin of the Plymouth district (see chapter 10, $F_1$ folds), and the northerly vergence of deformation structures near its northern boundary (Isaac et al., 1982) suggests that the geometry of the Tavy Basin was that of a full graben.

The deposits within the Tavy Basin represent relatively quiet-water deposition of mudstone and siltstone with sporadic incursions of turbiditic sandstone. A packet of such beds of sandstone composed of quartz-rich litharenite (Jones, 1993) constitutes the Trehills Sandstone Member. There are a few palaeocurrent indicators in the member which suggest flow towards the south-east, but within the main body of the Tavy Basin there is no good evidence for the transport direction of the argillaceous sediment. The grey-green mudstone and siltstone which forms the dominant lithology in the basin was deposited synchronously with purple and green mudstone and dark grey mudstone in the adjacent South Devon Basin and probably, therefore, had a different source (see p.66). The general uniformity of colour does not support bed by bed reduction, as a consequence of an overall greater proportion of organic detritus, as compared with the South Devon Basin. The lack of significant interdigitated dark grey mudstone (representing 'background' deposition from suspension) until late in the Famennian suggests that the deposition of the grey-green rocks of the Tavy Formation was relatively rapid but declined as the basin filled. Sparse purple and green mudstone, forming part of the Torpoint Formation, was deposited sporadically in the Tavy Basin, mainly during the uppermost Famennian, and may represent an overflow from the more extensive purple and green argillite of the Torpoint Formation in the South Devon Basin, possibly suggesting that subsidence in this basin slowed at this time.

The characteristic features of the evolution of the Tavy Basin differ from those of the adjacent South Devon and

Culm basins in two principle respects; the relative rarity of contemporaneous volcanic rocks and the consequent absence of stratabound mineralisation.

## Culm Basin

The Culm Basin, lying to the north of the Tavy Basin (Figure 37) , began to develop in the Tournaisian. There is little, if any, overlap between the age of the preserved sedimentary rocks forming the infilling of the Tavy Basin and the Culm Basin in the district, but the deposition of the Culm rocks in a separate basin is inferred from several lines of evidence:

i    There is no evidence of any sedimentary continuity between the rocks of the Tavy and Culm basins and their contrasting lithological types suggests that they were separately sourced

ii   The only contact between rocks of the Culm and Tavy basins is tectonic

iii  The Carboniferous rocks exhibit lower grades of deformation and metamorphism than the Tavy Basin rocks. This is consistent with them being metamorphosed to the north of the Tavy Basin within a regional pattern of northward decreasing metamorphism (see p.109)

iv   The St Mellion Formation rocks, thrust southwards over the Tavy Basin (see p.108), appear to define the southern marginal facies of the Culm Basin

The structural relationships between and current extent of the Crackington and Bude formations which form the bulk of the Culm Basin fill indicate that the basin, in the Upper Carboniferous, had an east–west trend (Thomas, 1988). The opposed vergence of the deformation structures in the Culm rocks, northwards on the northern margin of the basin and southwards on the southern margin (Dearman, 1971), indicates that the basin geometry was generally symmetrical. The presence of Tournaisian

and Viséan conglomeratic mudstone and olistostromes (see Isaac et al., 1982) and contemporaneous rift volcanic rocks (Rice-Birchall and Floyd, 1988) within the Dinantian of the Culm Basin suggests that, at least in its early stages, subsidence of the basin was fault controlled and that it therefore had the form of a symmetric graben. The significant feature of the Dinantian volcanic rocks is that, like those of the alkaline basalts of the Looe, South Devon and Tavy basins, they have the geochemical signature of Ocean Island Basalt (OIB) (Merriman et al., 2000) which in continental settings indicates major extension. Thus the Culm Basin originated in the extensional regime, even if later it became a foreland basin (Hartley and Warr, 1990). After deformation the Culm Basin is some 50 km in width, considerably larger than the Devonian basins.

During the Tournaisian the St Mellion Formation was deposited as a pro-delta complex (Jones, 1993; Figure 37). South-east to north-west, and east to west current directions from cross-bedding and sole markings in the formation (Whiteley, 1983) are compatible with the Culm Basin having an east–west trend with a delta complex on its southern margin. The more central parts of this basin, represented by the Brendon Formation, were dominated by mudstone sedimentation from suspension together with sporadic turbidites which reached out into the basin from the St Mellion pro-delta complex. Sedimentation was restricted on the northern side of the basin (Thomas, 1988).

The Tournaisian and Viséan volcanic rocks of the the Tintagel Volcanic Group in the Culm Basin (Freshney et al., 1972) are representative of very extensive volcanism throughout the Rhenohercynian Zone. It was the probable cause of the high silica content in the basinal waters which promoted the formation of radiolarian chert and siliceous siltstone (see Ziegler, 1982) of the Newton Chert Formation, in areas of low sediment influx. The lack of chert of equivalent age in the St Mellion Formation may have been due to the relatively

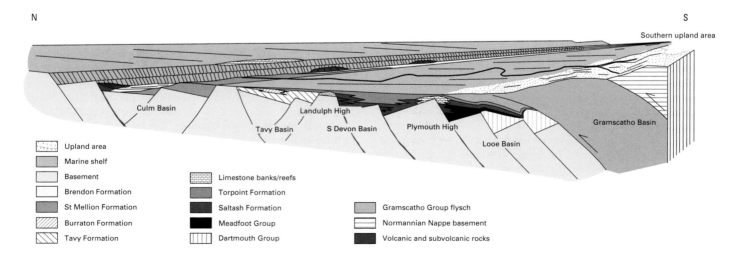

**Figure 37**   A synoptic view showing the development of the Culm Basin, the disposition of the main lithostratigraphical divisions, and the synchronous inversion of basins to the south.

high rates of clastic sediment deposition at the basin margin preventing chert accumulation. The OIB character of this volcanic episode is significant in indicating the continuity of the continental rifting regime (see p.89) into the Early Carboniferous.

Current directions indicate that the Culm Basin sediments were sourced from the south across the Tavy, South Devon and Looe basins. The sediment arriving in the Culm Basin was derived from a mixed continental source of basement rocks, low-grade metasedimentary rocks, chert and sandstone (Whiteley, 1983). The source was probably the Normannian Nappe (Holder and Leveridge, 1986b) and the Gramscatho Group in south Cornwall (Leveridge et al., 1990) which contain these lithologies, and which was undergoing inversion and up-thrusting at this time (Holder and Leveridge, 1986b). As the deformation front moved northwards with time (see below) inversion of the Looe and South Devon basins would have increased the source area and the amount of sediment transported into the Culm Basin and this is reflected in the northward progradation of proximal turbidites across the Culm Basin in late Viséan to Namurian times (Whiteley, 1983).

## VARISCAN DEFORMATION

### $D_1$ and $D_2$ deformation timing

The timing of the major compressional deformations in the district and throughout south-west England is largely reliant upon the radiometric dating of slates and cleaved mudstones by Dodson and Rex (1970). Their dates have since been amended by Warr et al. (1991) to take account of the decay constant revised by Steiger and Jager (1977). Modified dates, thought to be uplift ages because of a maximum resetting temperature of 250°C (Hunziker et al, 1986), are summarised in Figure 38. They reflect a general decrease northwards in the age of

southern crop of the Devonian slate, interrupted by a younger east–west belt between the Start Peninsula and Perranporth on the west coast of Cornwall, and younger metamorphic ages along the southern margin of the Culm Basin. This younging has been interpreted in terms of regional tectonic migration (e.g. Shackleton et al., 1982) and established by Leveridge and Holder (1985) and Holder and Leveridge (1986b) in south Cornwall where a dynamic stratigraphy of flysch and olistostrome sedimentation is the product of thrust-nappe migration northwards, associated with the obduction of the Lizard ophiolite. Warr (1993) and Warr et al. (1991) did not accept the migration concept but recognised two distinct cooling events within the east–west belt of rocks that includes those of the Plymouth district. The main 345 to 325 Ma (Viséan to mid-Namurian; Table 16) zone, which they attributed to $D_1$ inversion of the Trevone Basin, includes the Looe Basin, South Devon Basin and Tavy Basin. The zone to the north, with ages clustering in the 295 to 315 Ma (Westphalian to Stephanian) range, was assigned to a regional $D_2$ event associated with southerly thrusting along the southern margin of the Culm Basin. It is apparent that the Start–Perranporth ages also equate with this zone, and in the Perranporth area this corresponds with a zone of very strong $D_2$ transposition and transpressive fabrics (BGS unpublished data). The continuity of decreasing metamorphic ages associated with $D_1$ from south Cornwall through central Cornwall and south Devon then becomes apparent. It is also evident that $D_1$ continued northwards through the Culm Basin as sedimentation was active within that basin. Silesian rocks were probably largely deformed by $D_2$, but possibly in combination with $D_1$. Holder and Leveridge (1994) have suggested that the $D_2$ deformation, which is apparently synchronous throughout the European Variscides, was the result of continuing stress from collision to the south once the major extensional basins were closed.

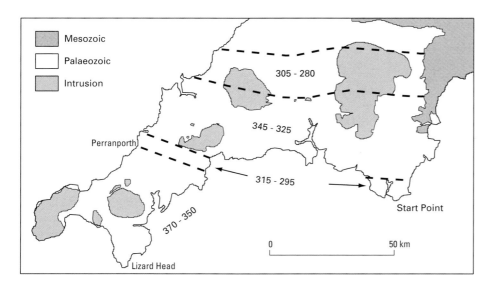

**Figure 38**   K-Ar metamorphic ages in south-west England (based on Dodson and Rex, 1970; amended by Warr et al., 1991).

**Table 16**   Devonian and Carboniferous chronostratigraphical ages (after Harland et al., 1989; Roberts et al., 1995).

| Period | Series | Sub-division | Stage | Age (Ma) |
|---|---|---|---|---|
| PERMIAN | | | | 252 Ma |
| | | | | 298 |
| CARBONIFEROUS | SILESIAN | | STEPHANIAN | |
| | | | | 306 |
| | | | WESTPHALIAN | |
| | | | | 313 |
| | | | NAMURIAN | |
| | | | | 328 |
| | DINANTIAN | Brigantian / Asbian / Holkerian / Arundian / Chadian | VISÉAN | |
| | | Courceyan | | 350 |
| | | | TOURNAISIAN | |
| | | | | 363 |
| DEVONIAN | UPPER | | FAMENNIAN | |
| | | | FRASNIAN | 377 |
| | MIDDLE | | GIVETIAN | |
| | | | EIFELIAN | 386 |
| | LOWER | | EMSIAN | |
| | | Siegenian | PRAGIAN | |
| | | Gedinnian | LOCHKOVIAN | 409 |
| SILURIAN | | | | 439 |

## Basin inversion and deformation

The extensional basins of the Plymouth district developed on the northern margin of the oceanised Gramscatho Basin. Migration of the $D_1$ compressive deformation northwards through the rifted terrain of east–west linear half-graben and full-graben basins and sub-basins caused those extensional basins to close up and invert. This process would be expected to proceed progressively with one basin or sub-basin closing and locking-up before stress transmission to its neighbour, to the north, commenced closure (Figure 39). In the Southern Domain, for example, the consequence of the fill of each of the Early Devonian sub-basins being expelled northwards over its neighbour is the translation of each bounding rift fault into a basal thrust, and the development of an associated major hanging-wall anticline. The juxtaposition of similar age deposits by the thrusting is a function of the extent of the inversion. At the more substantial high, forming a major basin margin, the development of short-cut footwall thrusting (Hayward and Graham, 1989) could be expected, being energetically more conservative than expelling the whole of the basin contents over the high. The detachment of the Plymouth Limestone Formation from its foundation by thrusting represents that process (see Figure 19). In the case of the Landulph High, it was overthrust by an imbricate of thin sheets from the rise slope suggesting that it was of smaller dimensions than the Plymouth High. The substantial overthrusting onto the Tavy Basin is also indicated by the passage of Saltash Formation divisions into the steep bounding Ludcote Fault which downfaults the overthrust sequences to the south. The geometry and facing of all structures in the Southern Domain is a function of the reactivation of southerly inclined, former extensional faults. In the cases of the Tavy and Culm basins, which are interpreted to be full-graben complexes, the southerly vergence and facing of major structures represents a primary northerly inclination of basinal faults rather than backfolding during inversion uplift (compare Selwood, 1990). The northerly facing at the southern margin of the Tavy Basin is a probable consequence of the immediately preceding overthrusting from the south.

By late Viséan to earliest Namurian times the inversion of all of the sedimentary basins to the south of the Culm Basin probably generated an extensive continental upland area with the Culm Basin representing a foredeep basin on its northern boundary: a foreland basin developing from the original extensional basin.

The migrating $D_1$ deformation front did not reach the Culm Basin, and initiate its inversion, until the Namurian. The southern margin deposits, which were rooted some 20 km to the north of the Plymouth district, were thrust southwards out of the basin, with the development of a major southerly facing, overturned, antiformal fold. A proportion of the Namurian and all of the Westphalian rocks of that basin had not been deposited at that time. Uplift and sedimentation must therefore have been concurrent, with the latter occurring in increasingly shallow water. The $D_2$

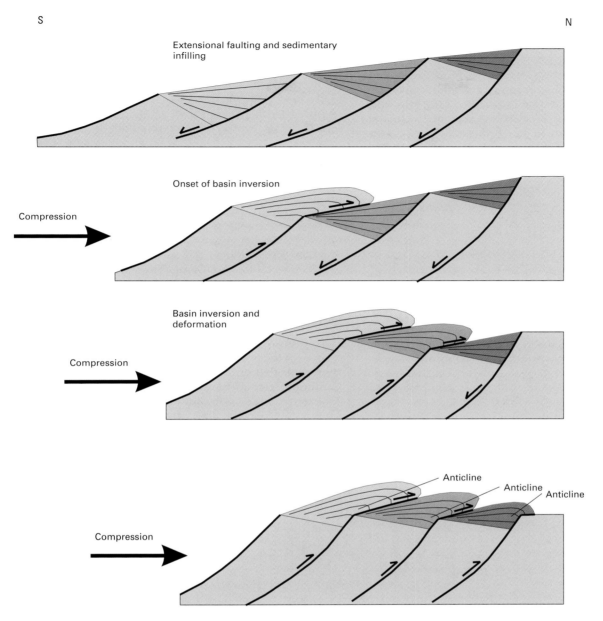

S

N

Extensional faulting and sedimentary infilling

Onset of basin inversion

Compression

Basin inversion and deformation

Compression

Compression

Anticline

Anticline

Anticline

**Figure 39**   Diagram illustrating sequential development of a stack of thrust anticlines by the compression and inversion of the sedimentary infills of half-graben basins. The presence of such regional-scale thrust anticlines in the district, without corresponding regional synclines, appears to be a natural consequence of the Devonian sedimentary rocks having been deposited in half-grabens.

deformation, at the end of the Westphalian, effected completion of the uplift and deformation of all the basin deposits. As the $D_2$ north-west–south-east maximum principal stress was oblique to the east–west trend of the primary basin, structures in the Upper Carboniferous rocks and on the northern margin of the basin reflect transpressive deformation (e.g. Andrews, 1993). This major cumulative uplift and southerly out-thrusting of the southern part of the basin fill, and the subsequent gravitational collapse of the edifice produced, is interpreted to be responsible for the Culm

Basin deposits in the Plymouth district. Most of the Carboniferous rocks between the Plymouth area and the Culm Basin are inverted (Seago and Chapman, 1988; unpublished BGS data) in thrust imbricates and duplexes. These cannot be explained by simple over-folding and require the delamination of the major fold at the southern margin of the basin during gravitational collapse (Figure 32). The $D_2$ deformation to the south, apart from causing the development of folds, also produced the out-of-sequence thrusting and reactivation of the strike-slip faults that had developed to the south during $D_1$, probably when basins had locked-up. Evidence of that reactivation is the compartmentalisation of the strong development of $D_2$ structures south of Liskeard by the Portwrinkle Fault which also limits the extensive flat overthrusting of the Staddon Formation present in the west. The rotation of earlier structures between the Portwrinkle and Portnadler faults appears to be linked to $D_2$ and the development of the major virgation of the Variscan of north-western Europe (see Holder and Leveridge, 1994). The $D_3$ southward-verging structures of this district and elsewhere are possible correlatives of the southerly directed thrusting of the Culm Basin rocks during the collapse of the uplifted basin and the early phase of late 'extensional' movements on $D_1$ and $D_2$ thrusts.

### Late Carboniferous–Permian extension

In south-west England, the latest stages of the Carboniferous and the Permian are marked by a number of tectonic events related to a post-Variscan phase of regional extension (Holder and Leveridge, 1994).

The onset of the uplift and extension is marked within the district by the intrusion of lamprophyric magma between 298 and 290 Ma (Rundle, 1980) into east–west-trending fractures. The intrusion of the Dartmoor and Bodmin granites between 290 and 280 Ma (see Darbyshire and Shepherd, 1985; Chen et al., 1993; Chesley et al., 1993) is linked to extension and pressure release leading to melting at the lower crust/mantle boundary (compare Watson et al., 1984). Related acid volcanic activity of the Kingsand Rhyolite Formation along the district's main north-west–south-east strike-slip faults, including the associated high-level elvan intrusion, and the intrusion of similar felsic magma into east–west fractures within and to the north of the district, is compatible with north–south extension at the time. Similarly the formation of the east–west-trending, main stage, mineral veins, dated at $280 \pm 20$ Ma by Moorbath (1962), indicates that the crust was undergoing north–south extension.

The magnitude of this latest Carboniferous to Permian extension is indicated by the formation of sedimentary basins, with several kilometres thickness of Permian sediments, above extensionally reactivated Variscan thrusts in Plymouth Bay (Evans, 1990); the Haig Fras, Melville and St Marys basins in the South-west Approaches (Hillis and Chapman, 1992) and the Crediton Trough in Devon (Durrance, 1985).

Within the district, a number of gently dipping thrusts exhibit significant extensional reactivation, most of which can be ascribed to this period of Permian extension. Within the Torpoint and Saltash formations proved in boreholes within the city of Plymouth, thrust faults display substantial hanging-wall breccias that were generated during extensional fault movements. At Warleigh Point [450 612] the thrust separating the Saltash Formation from those of the underlying Torpoint Formation displays a post-thrusting extensional offset. Thrusts within the Bovisand Formation between Tregantle Cliff [388 526] and Blackball Cliff [3714 5348] also display substantial extensional offsets (Figure 29).

### POST-VARISCAN TECTONICS

Rocks of Post-Permian to Quaternary age have not been recorded in the district, and so the direct evidence for post-Permian tectonic activity is limited. Farther east, the preservation of Cretaceous rocks on high ground around Kingsteignton (Selwood et al., 1984) suggests that the period between the Permian and Cretaceous was marked by general subsidence. Palaeogene deposits occupying the low-lying ground of the Bovey Basin in the same area (Selwood et al., 1984), together with a regular and continuous series of sea-level retreat features mapped in the Plymouth district during this survey, indicate the progressive emergence of south-west England from Cretaceous to Recent times. These tectonic events are summarised in Table 17.

### Events in the Plymouth district

The rotation of blocks between the Rame Fault, Portwrinkle Fault and Hoodney Cove Fault at Portwrinkle [360 537] indicates at least two periods of movement along the faults after the main dextral displacement and $D_2$ rotation during the Variscan orogeny. These post-Variscan movements appear to have been caused by a period of north–south extension followed by another period of north–south compression. The post-Permian age of at least some of the displacements along the Portwrinkle/Rame fault is demonstrated by the brecciation of Permian lava at Withnoe [404 518]. Similar Permian lava at Cawsand [430 508] has been brecciated and downfaulted on the north-west-trending Cawsand Fault suggesting a period of extensional faulting, possibly caused by north-east–south-west extensional tectonics.

The sinistral fault displacements on the Portwrinkle Fault could have been caused during north–south extension during the Permian, the Late Triassic to Jurassic or the early Palaeogene. Because the subsequent movement on the same fault appears to have been related to north–south compression the sinistral fault displacement appears limited to the Permian or Late Triassic to Jurassic extensional events. The subsequent dextral displacements of the Portwrinkle Fault are related to north–south compression which may be either of Mid Jurassic to Early Cretaceous or most probably of late Palaeogene/Neogene age.

There is no direct evidence within the district to indicate that displacement of Variscan thrusts occurred during post-Permian tectonism. However, the formation of the Wessex Basin by extensional reactivation of Variscan thrust faults and the closure of the Cainozoic Basins in south-west England by compressive reactivation of these thrusts suggests that at least some of the Variscan thrusts in the district may have had a long history of reactivation throughout the Mesozoic and Cainozoic.

**Table 17**   Summary of post-Variscan tectonic events likely to have affected south-west England.

| Age | Stress orientation | Event | References |
|---|---|---|---|
| Early to mid Triassic | E–W extension | Extension of Cheshire Basin in northern England north–south Pb–Ag–Zn–Sb mineral veins sourced by basinal brines | Evans et al., 1993 Scrivener et al., 1994 |
| Late Triassic and Jurassic | N–S extension | Formation of Wessex Basin by extensional reactivation of Variscan E–W faults | Chadwick, 1985, 1993 Bristow et al., 1991 |
| Late Jurassic to early Cretaceous | N–S compression | Basin inversion of Wessex Basin by compressive reactivation of E–W Variscan faults and dextral reactivation of NW–SE Variscan wrench faults Triassic to Early Cretaceous unconformity in Plymouth Bay, South Celtic Sea and Melville basins offshore | Bristow et al., 1991 Hillis and Chapman, 1992 Barton etal., 1993 Chadwick, 1993 Evans, 1990 |
| Cainozoic (Early to mid 'Tertiary') | NE–SW extension | Formation of Hampshire Basin and deposition of Palaeocene to Oligocene sediments; Palaeocene to Late Eocene sedimentation in the Western Approaches Melville Basin; formation of Petrockstow and Bovey basins along the NW–SE Sticklepath Fault with deposition of Eocene to Oligocene sediments<br><br>Vertical displacements of faults bounding the Bovey and Petrockstow basins indicate that these basins formed by sag across the Sticklepath Fault and not as pull-apart basins by dextral[†] or sinistral[‡] displacement<br><br>The lack of Variscan or early post-Variscan offsets along the Sticklepath Fault indicates the impossibility of producing the Cainozoic basins by the pull-apart mechanism | Edwards and Freshney, 1987 Plint, 1982 Evans, 1990 Freshney et al., 1979 Selwood et al., 1984 Hillis and Chapman, 1992 (cf. Vandycke et al., 1991) Bristow and Robson, 1994 Selwood et al., 1984 [†] Bristow et al., 1992 [‡] Holloway and Chadwick, 1986 see Emmons, 1969 Sylvester, 1988 |
| Cainozoic (Mid to late 'Tertiary') | N–S or NE–SW compression | Inversion of Hampshire Basin and Petrockstow and Bovey basins by compressive reactivation of E–W, and NW–SE faults<br><br>N–S compression causing dextral reorientation of bedding and cleavage strike along Portwrinkle Fault, affecting the post-Variscan, sinistral rotation of Variscan cleavage in Permian or Late Triassic and Jurassic events | Chadwick, 1993 Bristow and Hughes, 1971 Bristow and Robson, 1994<br><br>See text |
| Modern seismicity | NW–SE compression | Stress field measurements | Pinet et al., 1987 Evans and Brereton, 1990 Ahorner, 1975 Lisle, 1992 |

# INFORMATION SOURCES

Further geological information held by the British Geological Survey relevant to the Plymouth district is listed below. It includes published maps, memoirs and reports. Enquiries concerning geological data for the district should be addressed to the Manager, National Geological Records Centre, BGS, Keyworth. Geological advice for this area should be sought from the Programme Manager, Integrated Geoscience Surveys (South), BGS, Keyworth.

Other information sources include borehole records, mine plans, fossils, rock samples, thin sections, hydrogeological data and photographs. Searches of indexes to some of the collections can be made on the Geoscience Index system in British Geological Survey libraries. This is a developing computer-based system which carries out searches of indexes to collections and digital databases for specified geographical areas. The indexes which are available are listed below.

- Index of boreholes
- Topographical backdrop based on 1:250 000 scale maps
- Outlines of BGS maps at 1:50 000, 1:10 000, 1:10 560 and County Series maps
- Chronostratigraphical boundaries and areas from British Geological Survey 1:250 000 maps
- Geochemical sample locations on land
- Aeromagnetic and gravity data recording stations
- Land survey records

The BGS Catalogue of geological maps and books is available on request and BGS website can be accessed at: http://www.bgs.ac.uk

## MAPS

### Geological maps

*1:1 500 000*
Quaternary map of the United Kingdom (south) (1977)
Tectonic map of Britain, Ireland and adjacent areas (1996)

*1:1 000 000*
Pre-Permian geology of the United Kingdom (1985)

*1:625 000*
Solid geology of the United Kingdom (south) (1979)

*1:250 000*
Land's End 50N 06W Solid geology, 1985
Land's End 50N 06W Sea-bed sediments and Quaternary geology, 1987

*1:50 000 and 1:63 360*
Sheet 335 and 336 Trevose Head and Camelford, solid and drift, 1994
Sheet 337 Tavistock, solid and drift, 1994, provisional
Sheet 338 Dartmoor Forest, solid and drift, 1995, provisional
Sheet 347 Bodmin, solid and drift, 1982, reprinted with amendments
Sheet 348 Plymouth, solid and drift, 1998
Sheet 339 Ivybridge, solid and drift, 1974, (reprinted at 1:50 000 scale).
Sheet 353 and 354 Mevagissey, solid and drift, 1999.

*1:10 000 and 1:10 560*
The district was resurveyed between 1893 and 1908 at the six inch scale by G Barrow, D A MacAlister and W A E Ussher and published in 1907.

The maps at 1:10 000 scale covering the 1:50 000 Series Sheet 348 Plymouth are listed below, together with the surveyor's initials and the dates of the survey. The surveyors were C M Barton, A J J Goode, M T Holder, B E Leveridge and B J Williams.

The maps are not published but are available for consultation in the Library, British Geological Survey, Keyworth, and also at BGS Exeter Office and London Information Office in the Natural History Museum.

Dyeline copies may be purchased from the Sales Desk in Keyworth.

| | | |
|---|---|---|
| SX 25 NW | BEL, MTH, AJJG | 1992–1994 |
| SX 25 NE | AJJG | 1993–1994 |
| SX 25 SW | MTH | 1993–1994 |
| SX 25 SE | AJJG | 1992 |
| SX 26 NW* | MTH | 1993–1994 |
| SX 26 NE* | MTH | 1993–1994 |
| SX 26 SW | BEL, AJJG, MTH | 1994 |
| SX 26 SE | CMB additional information MTH | 1993, 1995 |
| SX 35 NW | CMB additional information AJJG | 1991–1992, 1995 |
| SX 35 NE | AJJG | 1991 |
| SX 35 SW | CMB additional information BEL | 1991, 1995 |
| SX 35 SE | BEL | 1992 |
| SX 36 NW* | MTH | 1987, 1993 |
| SX 36 NE* | MTH | 1987, 1991 |
| SX 36 SW | CMB additional information MTH | 1991, 1992, 1995 |
| SX 36 SE | MTH, BEL | 1986–1992 |
| SX 44 NW | BEL | 1986 |
| SX 45 NW | BEL, AJJG | 1987–1989 |
| SX 45 NE | BEL, AJJG | 1989–1991 |
| SX 45 SW | BEL, AJJG | 1986–1990 |
| SX 45 SE* | BEL, BJW, AJJG additional information MTH | 1986–1990, 1994 |
| SX 46 NW* | MTH, AJJG | 1987–1991 |
| SX 46 NE* | AJJG | 1992 |
| SX 46 SW | MTH, BEL, AJJG | 1987–1991 |
| SX 46 SE* | AJJG | 1986–1987 |

\* **Denotes sheets surveyed in part**

### Hydrogeological map

*1:625 000*
Sheet 18 (England and Wales), 1977

### Geochemistry maps

*1:625 000*
Methane, carbon dioxide and oil susceptibility, Great Britain — north and south, 1995
Radon potential based on solid geology, Great Britain — north and south, (in Appleton and Ball, 1995)

Distribution of areas with above national average background concentrations of potentially harmful elements (As, Cd, Cu, Pb and Zn), Great Britain — north and south, 1995

## Geophysical maps

*1:1 500 000*
Colour shaded relief gravity anomaly map of Britain, Ireland and adjacent areas (1996)
Colour shaded relief magnetic anomaly map of Britain, Ireland and adjacent areas (1996)

*1:250 000*
50N 06W   Land's End, Aeromagnetic anomaly, 1977
50N 06W   Land's End, Bouguer gravity anomaly, 1975

*1:50 000*
Geophysical information maps; these plot-on-demand maps are available which summarise graphically the publicly available geo-physical information held for the sheet in the BGS databases.

Features include:
- Regional gravity data — Bouguer anomaly contours and location of observations.
- Regional aeromagnetic data — total field anomaly contours and location of digitised data points along flight lines.
- Gravity and magnetic fields plotted on the same base map at 1:50 000 scale to show correlation between anomalies.
- Separate colour contour plots of gravity and magnetic fields at 1:125 000 scale for easy visualisation of important anomalies.
- Location of local geophysical surveys.
- Location of public domain seismic reflection and refraction surveys.
- Location of deep boreholes and those with geophysical logs.

## Minerals

*1:1 000 000*
Industrial minerals resources map of Britain (1996)

## BOOKS AND REPORTS

Memoirs, books, reports and papers relevant to the Plymouth district arranged by topic. Many are either out of print or are not widely available, but may be consulted at BGS and other libraries.

### British Regional Geology

South-west England, 4th Edition

The metalliferous mining region of south-west England, (Dines, H G, 1956; reprinted with addenda and corrigenda 1994)
Special report on the Mineral Resources of Great Britain. Vol. 21. Lead, silver-lead and zinc ores of Cornwall, Devon and Somerset, 1921 (out of print).

### United Kingdom Offshore Regional Report

The geology of the western English Channel and its western approaches, 1990

### Memoirs

Trevose Head and Camelford (Sheet 335, 336), 1996
Tavistock and Launceston (Sheet 337), 1911 (out of print)
Dartmoor (Sheet 338), 1912 (out of print)
Bodmin and St Austell (Sheet 347), 1909 (out of print)
Ivybridge and Modbury (Sheet 349), 1912 (out of print)
Mevagissey (Sheet 353), 1907 (out of print)
Kingsbridge and Salcombe (Sheet 355, 356), 1904 (out of print)

### Technical Reports

BARTON, C M, GOODE, A J J, and LEVERIDGE, B E. 1993. Geology of the St Germans district (Cornwall) 1:10 000 sheets SX 35 NW, SX 35 SW, SX 35 NE and SX 35 SE. *British Geological Survey Technical Report*, WA/93/93

BARTON, C M. 1994. Geology of the Liskeard district (Cornwall). 1:10 000 sheet SX 26 SE. *British Geological Survey Technical Report*, WA/94/30

GOODE, A J J, AND LEVERIDGE, B E, (contribution from R W O'B KNOX). 1993. Geological notes and details for 1:10 000 sheets SW 86 NW (combined, part) SW 87 NE, SW & SE, SW 96 NW (part) and NE (part) and SW 97 NW (part) and SW (part). (Trevose Head and St Breock Downs). *British Geological Survey Technical Report*, WA/93/93

### Biostratigraphy Reports

DEAN, A. 1993. Report on samples processed for palynomorphs 1/10/87–1/7/91. *British Geological Survey Technical Report*, WH/93/197R

DEAN, M T. 1993. Interim report, conodont biostratigraphic control for the Devonian rocks outcropping in the vicinity of Plymouth. *British Geological Survey Technical Report*, WH/93/73R

DEAN, M T. 1993. Interim report: conodont biostratigraphy of the Plymouth Limestone. Supplement to BGS Technical Report WH/93/73R. *British Geological Survey Technical Report*, WH/93/251R

DEAN, M T. 1994. Conodont biostratigraphic control for the Devonian rocks outcropping at Neal Point, Tamar estuary, South Devon. *British Geological Survey Technical Report*, WH/94/169R

DEAN, M T. 1994. Conodont biostratigraphic control for the Devonian rocks outcropping at Neal Point, Tamar estuary, South Devon. Addendum to BGS Technical Report WH/94/169R. *British Geological Survey Technical Report*, WH/94/199R

DEAN, M T. 1994. Interim report: conodont biostratigraphic control for the Devonian rocks outcropping at Plymouth Sound and southeast Cornwall. *British Geological Survey Technical Report*, WH/94/258R

DEAN, M T. 1994. A preliminary conodont based palaeoecological analysis of lithofacies of the Plymouth Limestone. *British Geological Survey Technical Report*, WH/94/80R

DEAN, M T. 1995. Interim report: conodont biostratigraphical control for Devonian rocks outcropping near Liskeard and Polbathic, southeast Cornwall. *British Geological Survey Technical Report*, WH/95/127R

FOREY, P. 1992. Devonian fish remains from 'The Long Stone' Cornwall. *British Geological Survey Technical Report*, WH/92/59R

HOUSE, M R. 1991. Report on goniatites from N Cornwall. *British Geological Survey Technical Report*, WH/91/170R

MCNESTRY, A. 1993. Devonian palynology of 8 samples from Torpoint and Tamar. *British Geological Survey Technical Report*, WH/93/333R

McNestry, A. 1993.   Palynological report on ?Devonian–Carboniferous samples from 1:50 000 sheets 347 and 348. *British Geological Survey Technical Report*, WH/93/134R

McNestry, A. 1993.   Palynology report on 11 samples from 1:50K Sheet 348.   *British Geological Survey Technical Report*, WH/93/334R

McNestry, A. 1994.   The palynology report of 2 samples from the Devonian of 1:50K Sheet 355.   *British Geological Survey Technical Report*, WH/94/145R

McNestry, A. 1994.   A summary of the palynostratigraphy and general biostratigraphy of 1:50K sheets 348, 349 and 355 Plymouth–Tamar area, Devon.   *British Geological Survey Technical Report*, WH/94/146R

Molyneux, S G. 1990.   Palynological analysis of Devonian samples for 1:10 000 sheet SX 45 SE.   *British Geological Survey Technical Report*, WH/90/79R

Molyneux, S G. 1990.   Palynological results from Brown Stone BH No.4, Devon.   *British Geological Survey Technical Report*, WH/90/229R

Molyneux, S G. 1990.   Palynological investigation of samples from 1:10 000 sheet SX 46 SW.   *British Geological Survey Technical Report*, WH/90/267R

Molyneux, S G. 1990.   Palynological investigation of samples from 1:10 000 sheet SX 45 NW.   *British Geological Survey Technical Report*, WH/90/268R

Molyneux, S G, and Owens, B. 1990.   Spores and acritarchs from samples of the Katebrook Slates, Devon.   *British Geological Survey Technical Report*, WH/90/343R

Molyneux, S G. 1991.   Palynological results from the Trevose Slate Formation, North Cornwall.   *British Geological Survey Technical Report*, WH/91/71R

Owens, B. 1988.   Palynological investigation of Devonian sediments from various localities on 1:50 000 sheet 335. *British Geological Survey Technical Report*, WH/88/384R

Owens, B. 1988.   Palynological report on ?Devonian samples from various localities on 1:50 000 sheet 348.   *British Geological Survey Technical Report*, WH/88/386R

Owens, B, McNestry, A M, and Turner, N.   1993. Palynological report on ?Devonian Carboniferous samples from various localities on 1:50 000 sheet 348.   *British Geological Survey Technical Report*, WH/93/123R

Rushton, A W A. 1990.   A fossil fish from Rame Head, Plymouth.   *British Geological Survey Technical Report*, WH/90/169R

Turner, N. 1992.   Palynological investigation to establish biostratigraphic control for the Devonian rocks outcropping in the vicinity of the lower Tamar estuary.   *British Geological Survey Technical Report*, WH/92/151R

Turner, N. 1993.   A new approach to processing low-grade metamorphic argillaceous rocks of Devonian age from the Tamar estuary and Plymouth districts.   *British Geological Survey Technical Report*, WH/93/15R

Turner, N. 1993.   Palynological report on a rock sample from 190m NNE Neal Point, west bank of the Tamar estuary.   *British Geological Survey Technical Report*, WH/93/75R

Wilkinson, I P. 1987.   Frasnian goniatites from Warren Point, near Plymouth.   *British Geological Survey Technical Report*, WH/87/253R

Wilkinson, I P. 1987.   Devonian Entomozoacean Ostracoda and Cricoconarida in the neighbourhood around Plymouth. *British Geological Survey Technical Report*, PD/87/374R

Wilkinson, I P. 1990.   Entomozoacean Ostracoda and Cricoconarida from the Upper Devonian Purple Green Slates of the Tamar estuary.   *British Geological Survey Technical Report*, WH/90/155R

Wilkinson, I P. 1990.   A re-examination of Ussher's Entomozoacean Ostracoda and Criconarida from the Upper Devonian Purple and Green Slates of the Tamar estuary. *British Geological Survey Technical Report*, WH/90/156R

Wilkinson, I P. 1993.   Calcareous faunas from Skinham Farm near Carkeel.   *British Geological Survey Technical Report*, WH/93/118R

Wilkinson, I P. 1993.   Entomozoacean Ostracods from the Grove–Markham (Markwell) area.   *British Geological Survey Technical Report*, WH/93/125R

Wilkinson, I P. 1993.   Microfauna from a suite of samples from 1-inch sheet 348.   *British Geological Survey Technical Report*, WH/93/146R

Wilkinson, I P. 1993.   A summary of the distribution of Devonian Entomozoacean Ostracoda from the Tamar estuary. WH/93/157R

**Engineering geology**

Fenwick, S M. 1993.   Engineering hazard assessment of planar discontinuities and karst in the Plymouth area.   *Internal*

Gostelow, T P. 1993.   Engineering geology of the Torpoint and Saltash formations, Plymouth.   *Internal*

**Geophysics**

Mould, A S. 1981.   Cruise report on project 74/03: A regional geophysical survey in the Plymouth Bay area, W. English Channel.   WB/MG/81/114

Rollin, K E. 1980.   Geophysical exploration for barytes near Ermington, South Devon.   *Applied Geophysics Unit Mineral Reconnaissance Report*: draft.

Shelton, A W. 1987.   The structure of the Plymouth limestones.   Unpublished report for BGS by University of Keels.

Smith, I F. 1994.   An assessment of detailed magnetic profiles on the Plymouth (348) Sheet, North of Liskeard.   PN/94/1

**Metamorphism**

Prior, S V, Merriman, R J. 1993.   White mica (illite) crystallinity of slates from the Plymouth district, SW England.   WG/93/12

Merriman, R J, Kemp, S J, Warr, L N, and Prior, S V. 1996. Regional low grade metamorphism in the Plymouth district, 1:50 000 Geological Sheet 348.   WG/96/9

**Mineralisation**

Beer, K E, et al. 1981.   Metalliferous mineralisation near Lutton, Ivybridge, Devon.   MRP 41

Cameron, D G. 1989.   Radon and thoron in soil gas traverses, South Hams, Devon.   WI/89/6/C

Jones, R C. 1981.   Reconnaissance geochemical maps of parts of south Devon and Cornwall.   MRP 44

Leake, R C, et al. 1985.   Volcanogenic and exhalative mineralisation within Devonian rocks of the South Hams district of Devon.   MRP 79

Leake, R C, et al. 1988.   Exploration for gold between the lower valleys of the Erme and Avon in the South Hams district of Devon.   MRP 98

Leake, R C et al. 1990. Gold and platinum group minerals in drainage between the River Erme and Plymouth Sound, S. Devon. WF/90/3

Leake, R C et al. 1991. Internal structure of Au Pd Pt grains from S. Devon, England, in relation to low-temperature transport and deposition. 01577

Leake, R C, and Norton, G E. 1993. Mineralisation in the Middle Devonian volcanic belt and associated rocks of S. Devon. WF/93/6

Perez-Alvarez, M. 1993. Mineralogical study of Sb mineralisation from Devon. MPSR/93/11

Toombs, J M C. 1980. Results of a gravity survey of the south-west margin of Dartmoor, Devon. MRP 34

## Radon

Appleton, J D, and Ball, T K. 1995. Radon and background radioactivity from natural sources: characteristics, extent and relevance to planning and development in Great Britain. WP/95/2

## Sedimentology

Davies, J R. 1987. Reconnaissance study of the Plymouth Limestone.

Humphreys, B, and Knox, R W O' B. 1987. Investigations into the sedimentology of Devonian sediments of the Plymouth area. SRG/87/14

Humphreys, B, and Smith, S A. 1988. The sedimentology of a regressive-transgressive sequence: the Middle Devonian Meadfoot Beds, Staddon Grits and Jennycliff Slates, Plymouth Sound. WH/88/256C

Jones, N S. 1992. Lacustrine and distal fan sediments and processes, Lower Devonian Dartmouth Group, SE Cornwall. WH/92/111R

Jones, N S. 1993. Sedimentology of Devonian and Carboniferous strata from the Plymouth Area, SW England. WH/93/77/R

Jones, N S. 1995. Shallow marine sedimentation in the Lower Devonian Meadfoot Group (Bovisand Formation) from the Whitsand Bay area. WH/95/210R

Scrivener, R C. 1984. Report on the geology of Moorcroft quarry (Elburton, Plymouth)

Scrivener, R C. 1985. Report on the geology of Pomphlett and Wixenford quarries (Plymstock, Plymouth).

Smith, S A. 1986. The sedimentology of the Dartmouth Beds of the Plymouth area: a reconnaissance study.

Smith, S A. 1988. Sedimentological aspects of the Dartmouth Group: Crownhill Bay to Wembury and Whitesand Bay to Rame. WH/88/196/C

Smith, S A, and Humphreys, B. 1989. The sedimentology and depositional setting of the Dartmouth Group, Bigbury Bay, S Devon. WH/89/345R

## MATERIAL COLLECTIONS

### Geological Survey photographs

Copies of the photographs that appear in this memoir are deposited for reference in the British Geological Survey library, Keyworth. Colour or black and white prints and transparencies can be supplied at a fixed tariff.

In addition, black and white photographs taken in 1904/1905 are held in BGS archives. The photographs show lavas and sedimentary rocks exposed in quarries and coastal sections; these may also be inspected at the library, Keyworth.

### Petrology

The petrological collections for the Plymouth district consist of samples and thin sections of the igneous, sedimentary and metamorphic rocks in the area and include some mineral sections. There are approximately 520 thin sections of igneous, sedimentary and metamorphic rocks and minerals. The collections are indexed on the basis of the one-inch geological maps but much of the early part of the collection cannot, at the moment, be searched by National Grid Reference.

### Borehole samples

At the time of going to press (2002) there are records of 1850 boreholes in BGS archives. Some of these are shallow boreholes drilled for site investigation; 84 are more than 30 m in depth.

Samples have been collected from core taken from boreholes; there are samples from nine boreholes in the district which are registered in the borehole collection.

### Palaeontology

The collections of biostratigraphical specimens are taken from surface and temporary exposures, and from boreholes throughout the Plymouth district. The collections are working collections and are used for reference. They are not at present on a computer database.

Macrofossils and micropalaeontological residues for samples collected from the district are held at BGS Keyworth. Enquiries concerning all macrofossil material should be directed to the Curator, Biostratigraphy Collections, BGS Keyworth.

## DATA

### BGS Lexicon of named rock unit definition

Definitions of the named rock units shown on BGS maps, including those shown on the 1:50 000 Series Sheet 348 Plymouth are held in the Lexicon database. This is available on Web Site http://www.bgs.ac.uk. Further information on the database can be obtained from the Lexicon Manager at BGS Keyworth.

### Geochemical data

Regional multi-element geochemical data are available for stream-sediment, stream-water and soil samples from the area. Enquiries should be directed to the Data Manager, G-BASE, BGS, Keyworth.

### Minerals

Directory of Mines and Quarries
United Kingdom Minerals Yearbook
MINGOL is a GIS based minerals information system, from which hard-copy and digital products tailored to individual clients' requirements can be obtained.

## ADDRESSES FOR DATA SOURCES

BGS hydrogeology enquiry service; wells, springs and water borehole records.
British Geological Survey, Hydrogeology Group
Maclean Building, Crowmarsh Gifford
Wallingford, Oxfordshire OX0 8BB
*Telephone*   01491 838800
*Fax*   01491 692345

British Geological Survey
Forde House
Park Five Business Centre
Sowton
Exeter
EX4 6BX
*Telephone*   01392 445271
*Fax*   01392 445371

London Information Office at the Natural History Museum
Earth Galleries
Exhibition Road, South Kensington
London SW7 2DE
*Telephone*   020–7589 4090
                      020–7942 5344/45
*Fax*   020–7584 8270

British Geological Survey (Headquarters)
Keyworth, Nottingham NG12 5GG
*Telephone*   0115 936 3100
*Fax*   0115 936 3200
*Web Site http://www.bgs.ac.uk.*

# REFERENCES

AHORNER, L. 1975. Present day stress field and seismotectonic block movements along major fault zones in Central Europe. *Tectonophysics*, Vol. 29, 233–249.

ALDERTON, D H M. 1978. Fluid inclusion data for lead-zinc ores from South-West England. *Transactions of the Institution of Mining and Metallurgy* (Section B), Vol. 87, B132–135.

ALDERTON, D H M. 1993. Mineralization associated with the Cornubian granite batholith. 270–354 in *Mineralization in the British Isles*. PATTRICK, R A D, and POLYA, D A (editors). (London: Chapman and Hall.)

AL-RAWI, F R J. 1980. A geophysical study of deep structures in south-west Britain. Unpublished PhD thesis, University of Wales.

ANDREWS, J R. 1993. Evidence for Variscan dextral transpression in the Pilton Shales, Croyde Bay, north Devon. *Proceedings of the Ussher Society*, Vol. 8, 198–199.

ATKINSON, K, BOULTER, M C, FRESHNEY, E C, WALSH, R T, and WILSON, A C. 1975. A revision of the geology of the St Agnes outlier, Cornwall (Abstract). *Proceedings of the Ussher Society*, Vol. 3, 286–287.

AUSTIN, R L, ORCHARD, M J, and STEWART, I J. 1985. Conodonts of the Devonian System from Great Britain. 93–166 in *A stratigraphical index of conodonts*. HIGGINS, A C, and AUSTIN, R L (editors). (Chichester: Ellis Horwood Limited.)

BADHAM, J P N. 1982. Strike-slip orogens — an explanation for the Hercynides. *Journal of the Geological Society of London*, Vol. 139, 493–504.

BANDEL, K, and BECKER, G. 1975. Ostracoden aus Palaozoischen-pelagischen der Karnischen Alpen (Silurium bis Unterkarbon). *Senckenbergiana lethaea*, Vol. 56, 1–283. [German]

BARD, J P, BURG, J P, MATTE, P, and RIBIERO, A. 1980. La Chaine Hercynienne d'Europe occidentale en termes de tectoniques des plaques. 90–111 in *Geologie de l'Europe*. COGNE, J, and SLANSKY, M (editors). (Paris: 26th International Geological Congress.). [French]

BARTON, C M. 1994. Geology of the Liskeard district (Cornwall). 1:10 000 Sheet SX26SE. *Technical Report of the British Geological Survey, Onshore Geology Series*, WA/94/30.

BARTON, C M, EVANS, D J, BRISTOW, C R, FRESHNEY, E C, and KIRBY, G A. 1998. Reactivation of relay ramps and structural evolution of the Mere Fault and Wardour Monocline, northern Wessex Basin. *Geological Magazine*, Vol. 135, 383–395.

BARTON, C M, GOODE, A J J, and LEVERIDGE, B E. 1993. Geology of the St Germans district (Cornwall). 1:10 000 sheets SX35NW, SX35SW, SX35NE and SX35SE. *Technical Report of the British Geological Survey, Onshore Geology Series*, WA/93/93.

BARTON, D B. 1964. *A historical survey of the mines and mineral railways of east Cornwall and west Devon*. (Truro: Truro Bookshop.) 102 pp.

BEAMISH, D, and SMITH, I F. 1994. A magnetotelluric survey southwards from the Bodmin Granite. *Technical Report of the British Geological Survey*, WN/94/04.

BEER, K E. 1988. Addenda and corrigenda to 'The Metalliferous mining region of south-west England'. H G DINES (1956). *Economic Memoir of the Geological Survey of Great Britain*.

BEHR, H J, ENGEL, W, and FRANKE, W. 1982. Variscan wildflysch and nappe tectonics in the Saxothuringian Zone (Northeast Bavaria, West Germany). *American Journal of Science*, Vol. 282, 1438–1470.

BRITISH GEOLOGICAL SURVEY. 1985a. Land's End. Sheet 50°N 06°W. Bouguer Anomaly. 1:250 000. (Southampton: Ordnance Survey for British Geological Survey.)

BRITISH GEOLOGICAL SURVEY. 1985b. Land's End. Sheet 50°N 06°W. Aeromagnetic Anomaly. 1:250 000. (Southampton: Ordnance Survey for British Geological Survey.)

BRITISH GEOLOGICAL SUVEY. 1985c. Land's End. Sheet 50°N 06°W. Solid Geology. 1:250 000. (Southampton: Ordnance Survey for British Geological Survey.)

BRITISH GEOLOGICAL SURVEY. 1987. Land's End. Sheet 50N 06W. Quaternary geology and sea-bed sediments. 1:250 000 UTM Series. (Southampton: Ordnance Survey for British Geological Survey.)

BRITISH GEOLOGICAL SURVEY. 1990. 1:10 000 series Geological Sheet, SX 64 NW.

BRITISH GEOLOGICAL SURVEY. 1996a. 1:10 000 Series Geological Sheet, SX 45 SE.

BRITISH GEOLOGICAL SURVEY. 1996b. 1:10 000 Series Geological Sheet, SX 45 NE.

BOGLI, A. 1980. *Karst hydrology and physical speleology*. (Berlin, Heidelberg: Springer-Verlag.)

BOOKER, F. 1967. *Industrial archaeology of the Tamar Valley*. (Newton Abbot: David and Charles.) 303 pp.

BOTT, M H P. 1992. Passive margins and their subsidence. *Journal of the Geological Society of London*, Vol. 149, 805–812.

BOTT, M H P, DAY, A A, and MASSON-SMITH, D. 1958. The geological interpretation of gravity and magnetic surveys in Devon and Cornwall. *Philosophical Transactions of the Royal Society of London*. Vol. 251A, 161–191.

BOTT, M H P, and SCOTT, P. 1964. Recent geophysical studies in south-west England. 25–44 in *Present views of some aspects of the geology of Cornwall and Devon*. HOSKING, K F G, and SHRIMPTON, G J (editors). (Penzance: Royal Geological Society of Cornwall.)

BOUMA, A H. 1962. *Sedimentology of some flysch deposits: A graphic approach to facies interpretation*. (Amsterdam: Elsevier.) 168 pp.

BOWEN, D Q. 1969. A new interpretation of the Pleistocene succession in the Bristol Channel area. *Proceedings of the Ussher Society*, Vol. 2, 86.

BOWEN, D Q. 1973. The Pleistocene succession of the Irish Sea. *Proceedings of the Geologists' Association*, Vol. 84, 249–272.

BOWEN, D Q. 1994. Late Cenozoic Wales and south-west England. *Proceedings of the Ussher Society*, Vol. 8, 209–213.

BRAITHWAITE, C J R. 1967. Carbonate environments in the Middle Devonian of south Devon, England. *Sedimentary Geology*, Vol. 1, 283–320.

BRAY, C J, and SPOONER, E.T.C. 1983. Sheeted vein Sn–W mineralization and greisenization associated with economic kaolinisation, Goonbarrow china clay pit, St Austell, Cornwall, England. *Economic Geology*, Vol. 78, 1064–1089.

BRISTOW, C M, and HUGHES, D E. 1971. A Tertiary thrust fault on the southern margin of the Bovey Basin. *Geological Magazine*, Vol. 108, 61–68.

BRISTOW, C M, PALMER, Q G, and PIRRIE, D. 1992. Palaeogene basin development: new evidence from the southern Petrockstow Basin, Devon. *Proceedings of the Ussher Society*, Vol. 8, 19–22.

BRISTOW, C M, and ROBSON, J L. 1994. Palaeocene basin development in Devon. *Proceedings of the Institute of Mining and Metallurgy*, Vol. 103B, 163–174.

BRISTOW, C R, FRESHNEY, E C, and PENN, I E. 1991. Geology of the country around Bournemouth. *Memoir of the British Geological Survey*, Sheet 329 (England and Wales).

BROOKS, M, DOODY, J J, and AL-RAWI, F R J. 1984. Major crustal reflectors beneath SW England. *Journal of the Geological Society of London*, Vol. 141, 97–103.

BULL, B. 1982. Mineralisation of an area around Tavistock, South-West England. Unpublished PhD thesis, University of Exeter.

BURTON, C J. 1972. Provincial affinities of Eifelian Phacopids (Trilobita) of South West England and Brittany. *Proceedings of the Ussher Society*, Vol. 2, 458–463.

BURTON, C J. 1974. A progress report on geological investigations in the Liskeard area of south east Cornwall (Abstract). *Proceedings of the Ussher Society*, Vol. 3, 44–48.

BURTON, C J, and CURRY, G B. 1985. Pelagic brachiopods from the Upper Devonian of East Cornwall. *Proceedings of the Ussher Society*, Vol. 6, 191–195.

BURTON, C J, and TANNER, P W G. 1986. The stratigraphy and structure of the Devonian rocks around Liskeard, east Cornwall, with regional implications. *Journal of the Geological Society of London*, Vol. 143, 95–105.

CHADWICK, R A. 1985. Permian, Mesozoic and Cenozoic structural evolution of England and Wales in relation to the principles of extension and inversion tectonics. 9–25 in *Atlas of onshore sedimentary basins in England and Wales*. WHITTAKER, A (editor). (London: Blackie.)

CHADWICK, R A. 1993. Aspects of basin inversion in southern Britain. *Journal of the Geological Society of London*, Vol. 150, 311–322.

CHAMPERNOWNE, A. 1881. Notes on a find of *Homalonotus* in the Red Beds at Torquay. *Geological Magazine*, Vol. 8, 487–488.

CHANDLER, P, and MCCALL, G J H. 1985. Stratigraphy and structure of the Plymouth–Plymstock area: a preliminary revision. *Proceedings of the Ussher Society*, Vol. 6, 253–257.

CHAPMAN, T J. 1983. A guide to the structure of the Lower to Middle Devonian Staddon Grits and Jennycliff Slates on the east side of Plymouth Sound, Devon. *Proceedings of the Ussher Society*, Vol. 5, 460–464.

CHAPMAN, T J, FRY, R L, and HEAVEY, P T. 1984. A structural cross-section through SW Devon. 113–118 in *Variscan tectonics of the North Atlantic Region*. HUTTON, D H W, and SANDERSON, D J (editors). (Oxford: Blackwell Scientific Publications.)

CHEN, Y, CLARK, A H, FARRAR, E, WASTENEYS, H A H P, HODGSON, M J, and BROMLEY, A V. 1993. Diachronous and independent histories of plutonism and mineralisation in the Cornubian Batholith, southwest England. *Journal of the Geological Society of London*, Vol. 150, 1183–1191.

CHESLEY, J T, HALLIDAY, A N, and SCRIVENER, R C. 1991. Samarium–Neodymium direct dating of fluorite mineralisation. *Science*, Vol. 252, 949–951.

CHESLEY, J T, HALLIDAY, A N, SNEE, L W, MEZGER, K, SHEPHERD, T J, and SCRIVENER, R C. 1993. Thermochronology of the Cornubian batholith in southwest England: implications for pluton emplacement and protracted hydrothermal mineralisation. *Geochimica et Cosmochimica Acta*, Vol. 57, 1817–1835.

CLAYTON, R E, SCRIVENER, R C, and STANLEY, C J. 1990. Mineralisation and preliminary fluid inclusion studies of lead-antimony mineralisation in north Cornwall. *Proceedings of the Ussher Society*, Vol. 7, 258–262.

CLEMMENSON, L B. 1979. Triassic lacustrine red-beds and palaeoclimate: the "Buntsandstein" of Helgoland and the Malmros Klint Member of East Greenland. *Geologische Rundschau*, Vol. 68, 748–774.

CLIFTON, H E. 1973. Pebble segregation and bed lenticularity in wave-worked versus alluvial gravel. *Sedimentology*, Vol. 20, 173–187.

CLIFTON, H E. 1976. Wave-formed sedimentary structures — a conceptual model. 126–148 in Beach and nearshore sedimentation. DAVIS, R A, and ETHERINGTON, R L (editors). *Special Publication of the Society of Economic Palaeontologists and Mineralogists*, No. 24.

COBBOLD, P R, and QUINQUIS, H. 1980. Development of sheath folds in shear regimes. *Journal of Structural Geology*, Vol. 2, 119–126.

CODRINGTON, T. 1898. On some rock-valleys in south Wales, Devon and Cornwall. *Quarterly Journal of the Geological Society of London*, Vol. 54, 251–278.

COLLINS, J H. 1904. Notes on the principal lead-bearing lodes of the west of England. *Transactions of the Royal Geological Society of Cornwall*, Vol.12, 683–718.

COLLINS, J H. 1912. Observations on the west of England mining region. *Transactions of the Royal Geological Society of Cornwall*, Vol.14, 683.

COLLINSON, J D, MARTINSON, O, BAKKEN, B, and KLOSTER, A. 1991. Early fill of the Western Irish Namurian Basin: a complex relationship between turbidites and deltas. *Basin Research*, Vol. 3, 223–242.

CORNWALL COUNTY COUNCIL. 1996. Minerals local plan, deposit version.

CORNWELL, J D. 1967a. Palaeomagnetism of the Exeter Lavas, Devonshire. *Geophysical Journal of the Royal Astronomical Society*, Vol. 12, 181–196.

CORNWELL, J D. 1967b. The magnetism of Lower Carboniferous rocks from the north-west border of the Dartmoor Granite, Devonshire. *Geophysical Journal of the Royal Astronomical Society*, Vol. 12, 381–403.

COSGROVE, M E, and ELLIOTT, M H. 1976. Supra-batholithic volcanism of the south-west England granites. *Proceedings of the Ussher Society*, Vol. 3, 391–401.

COWARD, M P, and MCCLAY, K R. 1983. Thrust tectonics of S Devon. *Journal of the Geological Society of London*, Vol. 140, 214–228.

COX, K G, BELL, J D, and PANKHURST, R J. 1979. *The interpretation of igneous rocks*. (London: George Allen and Unwin.)

CULLINGFORD, R A. 1982. The Quaternary. 249–288 in *The geology of Devon*. DURRANCE, E M, and LAMING, D J C (editors). (Exeter: University of Exeter.)

DARBYSHIRE, D P F. 1995. Late vein mineralisation in the Plymouth district. *NERC Isotope Geosciences Laboratory Report*, No. 68.

DARBYSHIRE, D P F, and SHEPHERD, T J. 1985. Chronology of granite magmatism and associated mineralization, SW England. *Journal of the Geological Society of London*, Vol. 142, 1159–1177.

DARBYSHIRE, D P F, and SHEPHERD, T J. 1987. Chronology of magmatism in south-west England: the minor intrusions. *Proceedings of the Ussher Society*, Vol. 6, 431–438.

DARBYSHIRE, D P F, and SHEPHERD, T J. 1994. Nd and Sr isotope constraints on the origin of the Cornubian batholith, SW England. *Journal of the Geological Society of London*, Vol. 151, 795–802.

DARNLEY, A G, ENGLISH, T H, SPRAKE, O, PREECE, E R, and AVERY, D. 1965. Ages of uraninite and coffinite from south-west England. *Mineralogical Magazine*, Vol. 34, 159–176.

DAVISON, E H. 1926. *Handbook of Cornish geology*. (Penzance: Royal Geological Society of Cornwall.) 103 pp.

DEAN, A. 1989a. Short paper: Palynomorphs from deformed low-grade metamorphic rocks: an SEM-based technique. *Journal of the Geological Society of London*, Vol. 146, 597–599.

DEAN, A. 1989b. A new assemblage of palynomorphs from the low-grade Upper Devonian metamorphic rocks of east Cornwall. *Proceedings of the Ussher Society*, Vol. 7, 180–182.

DEAN, A. 1991. Upper Palaeozoic palynomorphs from the low grade metamorphic rocks of Devon and Cornwall. Unpublished PhD thesis, University of Exeter.

DEAN, A. 1992. Report on samples processed for palynomorphs 1/10/87–1/7/91. *Report of the British Geological Survey, Biostratigraphy Series*, WH/93/197R.

DEAN, M T. 1993. Interim report: conodont biostratigraphic control for the Devonian rocks outcropping in the vicinity of Plymouth. *Technical Report of the British Geological Survey, Stratigraphy Series*, WH/93/73R.

DEAN, M T. 1994a. Interim report: conodont biostratigraphic control for the Devonian rocks outcropping at Plymouth Sound and south-east Cornwall. *Technical Report of the British Geological Survey, Stratigraphy Series*, WH/94/258R.

DEAN, M T. 1994b. Conodont biostratigraphic control for the Devonian rocks outcropping at Neal Point, Tamar Estuary, South Devon. *Technical Report of the British Geological Survey, Stratigraphy Series*, WH/94/169R.

DEAN, M T. 1995. Interim report: conodont biostratigraphic control for Devonian rocks outcropping near Liskeard and Polbathic, south-east Cornwall. *Technical Report of the British Geological Survey, Stratigraphy Series*, WH/95/127R.

DEARMAN, W R. 1963. Wrench faulting in Cornwall and south Devon. *Proceedings of the Geologists' Association*, Vol. 74, 265–287.

DEARMAN, W R. 1964. Refolded folds in the Dartmouth Slates at Portwrinkle, south Cornwall. *Proceedings of the Ussher Society*, Vol. 1, 79–81.

DEARMAN, W R. 1969. Tergiversate folds from south-west England. *Proceedings of the Ussher Society*, Vol. 2, 112–115.

DEARMAN, W R. 1971. A general view of the structure of Cornubia. *Proceedings of the Ussher Society*, Vol. 2, 220–236.

DEARMAN, W R. 1991. *Engineering geology mapping*. (London: Butterworth–Heineman.)

DEARMAN, W R, LEVERIDGE, B E, and TURNER, R G. 1969. Structural sequences and the ages of slates and phyllites from south-west England. *Proceedings of the Geological Society of London*, Vol. 1654, 41–45.

DE LA BECHE, H T. 1839. Report on the geology of Devon, Cornwall and west Somerset. *Memoir of the Geological Survey of Great Britain*.

DE RAAF, J F M, BOERSMA J R, and VAN GELDER, A. 1977. Wave generated structures and sequences from a shallow marine succession, Lower Carboniferous, County Cork, Ireland. *Sedimentology*, Vol. 39, 89–101.

DEWEY, H. 1925. The mineral zones of Cornwall. *Proceedings of the Geologists' Association*, Vol. 36, 107–135.

DINELEY, D L. 1966. The Dartmouth Beds of Bigbury Bay, south Devon. *Quarterly Journal of the Geological Society of London*, Vol. 122, 187–217.

DINES, H G. 1934. The lateral extent of the ore-shoots in the primary depth zones of Cornwall. *Transactions of the Royal Geological Society of Cornwall*, Vol. 16, 279–96.

DINES, H G. 1956. The metalliferous mining region of south-west England. *Economic Memoir of the Geological Survey of Great Britain*.

DODSON, M H, and REX, D C. 1971. Potassium-argon ages of slates and phyllites from south-west England. *Quarterly Journal of the Geological Society of London*, Vol. 141, 315–326.

DOODY, J J, and BROOKS, M. 1986. Seismic refraction investigation of the structural setting of the Lizard and Start complexes, SW England. *Journal of the Geological Society of London*, Vol. 143, 135–140.

DUBUISSON, G, MERCIER, J-C C, GIRARDEAU, J, and FRISON, J-Y. 1989. Evidence for a lost ocean in Variscan terranes of the western Massif Central, France. *Nature, London*, Vol. 337, 729–732.

DUNHAM, R J. 1962. Classification of carbonate rocks according to depositional texture. 108–121 in Classification of carbonate rocks. HAM, W E (editor). *Memoir of the American Association of Petroleum Geologists*, No. 1.

DURAND, S. 1977. Bretagne. *Guides Geologiques Regionaux*. (Paris: Masson.) [French]. 208 pp.

DURRANCE, E M. 1985. A possible major Variscan thrust along the southern margin of the Bude Formation, south-west England. *Proceedings of the Ussher Society*, Vol. 6, 173–179.

DURRANCE, E M, BROMLEY, A V, BRISTOW, C M, HEATH, M J, and PENMAN, J M. 1982. Hydrothermal circulation and post-magmatic changes in granites of south-west England. *Proceedings of the Ussher Society*, Vol. 5, 304–320.

EAGAR, R M C, BAINES, J G, COLLINSON, J D, HARDY, P G, OKOLO, S A, and POLLARD, J E. 1985. Trace fossil assemblages and their occurrence in Silesian (mid-Carboniferous) deltaic sediments of the central Pennine basin, England. 99–149 in Biogenic structures: their use in interpreting depositional environments. CURRAN, H A (editor). *Society of Economic Palaeontologists and Mineralogists, Special Publication*, No. 35.

EDDIES, R D, and REYNOLDS, J M. 1988. Seismic characteristics of buried rock-valleys in Plymouth Sound and the River Tamar. *Proceedings of the Ussher Society*, Vol. 7, 36–40.

EDMONDS, E A, McKEOWN, M C, and WILLIAMS, M. 1975. *British regional geology: south-west England*. (London: HMSO for Institute of Geological Sciences.)

EDMONDS, E A, WRIGHT, J E, BEER, K E, HAWKES, J R, WILLIAMS, M, FRESHNEY, E C, and FENNING, P J. 1968. Geology of the country around Okehampton. *Memoir of the Geological Survey of Great Britain*, Sheet 324 (England and Wales).

EDWARDS, R A, and FRESHNEY. 1987. Geology of the country around Southampton. *Memoir of the British Geological Survey*, Sheet 315 (England and Wales).

EDWARDS, R A, and SCRIVENER, R C. 1999. The geology of the country around Exeter. *Memoir of the British Geological Survey*, Sheet 325 (England and Wales).

EDWARDS, R P. 1976. Aspects of trace metal and ore distribution in Cornwall. *Transactions of the Institute of Mining and Metallurgy* (Section B: Applied Earth Sciences), Vol. 85, B83–90.

EMBRY, A F, and KLOVAN, J E. 1971. A late Devonian reef tract on northeastern Banks Island, Northwest Territories. *Bulletin of Canadian Petroleum Geology*, Vol. 19, 730–781.

EMMONS, R C. 1969. Strike-slip rupture patterns in sand models. *Tectonophysics,* Vol. 7, 71–87.

ENDS. 1992. *Dangerous substances in water — a practical guide.* Environmental Data Services Ltd. 43 pp.

ENGEL, W, FRANKE, W, and LANGENSTRASSEN, F. 1983. Palaeozoic sedimentation in the Northern Branch of the Mid-European Variscides — essay of an interpretation. 9–41 in *Intracontinental fold belts.* MARTIN, H, and EDER, F W (editors). (Berlin: Springer-Verlag.)

EVANS, C D R. 1990. The geology of the western English Channel and its western approaches. *United Kingdom offshore regional report.* (London: HMSO for the British Geological Survey.)

EVANS, D J, and BRERETON, N R. 1990. In situ crustal stress in the UK from borehole breakouts. 327–338 in Geological application of wireline logs. HIRST, A (editor). *Special Publication of the Geological Society of London*, No. 48.

EVANS, D J, REES, J G, and HOLLOWAY, S. 1993. The Permian to Jurassic stratigraphy and structural evolution of the central Cheshire Basin. *Journal of the Geological Society of London,* Vol. 150, 857–870.

EVANS, K M. 1981. A marine fauna from the Dartmouth Group (Lower Devonian) of Cornwall. *Geological Magazine,* Vol. 118, 517–523.

EVANS, K M. 1983. The marine Lower Devonian of the Plymouth area (Abstract). *Proceedings of the Ussher Society,* Vol. 5, 489.

FLOYD, P A. 1982. Chemical variation in Hercynian basalts relative to plate tectonics. *Journal of the Geological Society of London,* Vol. 139, 505–520.

FLOYD, P A. 1984. Geochemical characteristics and comparison of the Lizard Complex and the basaltic lavas within the Hercynian troughs of SW England. *Journal of the Geological Society of London,* Vol. 141, 61–70.

FOOKES, P G, REEVES, B J, and DEARMAN, W R. 1977. The design and construction of a rock slope in weathered slate at Fowey, SW England. *Geotechnique,* Vol. 27, 533–556.

FOSTER, C LE N. 1878. Remarks on the lode at Wheal Mary Ann, (Men)hemiot. *Transactions of the Royal Geological Society of Cornwall,* Vol. 9, 152–157.

FOX, H H. 1896. The radiolarian cherts of Cornwall. *Transactions of the Royal Geological Society of Cornwall,* Vol. 12, 39–70.

FRANKE, W. 1989. Tectonostratigraphic units in the Variscan belt of Central Europe. *Geological Society of America, Special Paper,* 230, 67–90.

FRANKE, W, and PAUL, J. 1980. Pelagic redbeds in the Devonian of Germany — deposition and diagenesis. *Sedimentary Geology,* Vol. 25, 231–256.

FRESHNEY, E C, BEER, K E, and WRIGHT, J E. 1979. Geology of the country around Chulmleigh. *Memoir of the British Geological Survey,* Sheet 309 (England and Wales).

FRESHNEY, E C, McKEOWN, M N, and WILLIAMS, M. 1972. The geology of the coast between Tintagel and Bude. *Memoir of the Geological Survey of Great Britain,* Sheet 322 (England and Wales).

FREY, M, DE CAPITANI, C, and LIOU, J G. 1991. A new petrographic grid for low-grade metabasites. *Journal of Metamorphic Geology,* Vol. 9, 497–509.

FRIEND, P F, HIRST, J P P, and NICHOLS, G J. 1986. Sandstone body structure and river processes in the Ebro Basin of Argon, Spain. *Cuadernos Geologia Iberia,* Vol. 10, 9–30.

GAPS. 1985. Environmental geology study of Plymouth–Plymstock. Unpublished report, Department of the Environment.

GARLAND, J, TUCKER, M E, and SCRUTTON, C T. 1996. Microfacies analysis and metre scale cyclicity in the Givetian back-reef sediments of SE Devon. *Proceedings of the Ussher Society,* Vol. 9, 31–36.

GEACH. 1936. An account of a subterranean cave at Stonehouse near Plymouth. *South Devon Monthly Museum Journal.*

GIBBS, A D. 1984. Structural evolution of extensional basin margins. *Journal of the Geological Society of London,* Vol. 141, 609–620.

GILES, J. 1865. On the metalliferous associations of the Liskeard rocks. *Transactions of the Royal Geological Society of Cornwall,* Vol. 7, 198–207.

GOLDRING, R. 1962. The bathyal lull: Upper Devonian and Lower Carboniferous sedimentation in the Variscan Geosyncline. 75–91 in *Some aspects of the Variscan Fold Belt.* COE, K (editor). (Manchester: University Press.)

GOLDRING, R, and LANGENSTRASSEN, F. 1979. Nearshore clastic facies. 81–98 *in* The Devonian System. SCRUTTON, C T, and BASSETT, M G (editors). *Special Report in Palaeontology, Palaeontological Association, London,* No. 23.

GOODAY, A J. 1973. Taxonomic and stratigraphic studies on Upper Devonian and Lower Carboniferous Entomozoidae and Rhomboentomozoidae (Ostracoda, ?Myodocopida) from Southwestern England. Unpublished PhD thesis, University of Exeter.

GOODAY, A J. 1974. Ostracod ages from purple and green slates around Plymouth. *Proceedings of the Ussher Society,* Vol. 3, 55–62.

GOODE, A J J, and LEVERIDGE, B E. 1991. Geological notes and details for 1:10 000 sheets SW 86 NW and NE (combined, part), SW 87 NE, SW and SE, SW 96 NW (part) and NE (part) and SW 97 NW (part) and SW (part). (Trevose Head and St Breock Downs). *Technical Report of the British Geological Survey, Onshore Series,* WA/92/11.

GOODE, A J J, MERRIMAN, R J, and DARBYSHIRE, D P F. 1987. Evidence of crystalline basement west of the Land's End granite, Cornwall: a reply. *Proceedings of the Geologists' Association,* Vol. 98, 272.

GOODE, A J J, and TAYLOR, R T. 1988. Geology of the country around Penzance. *Memoir of the British Geological Survey,* Sheets 351 and 358 (England and Wales).

HALLIDAY, A N. 1980. The timing of early and main stage ore mineralisation in southwest Cornwall. *Economic Geology,* Vol. 75, 752–759.

HAMILTON JENKIN, A K. 1967. *Mines and miners of Cornwall, Volume 14, St Austell to Saltash.* (Bracknell: Town and Country Press.)

HAMILTON JENKIN, A K. 1974. *Mines of Devon.* (Newton Abbot: David and Charles.) 154 pp.

HARLAND, W B, ARMSTRONG, R L, COX, L V, CRAIG, L E, SMITH, A G, and SMITH, D G. 1989. *A geologic time scale.* (Cambridge: Cambridge University Press.)

HARTLEY, A J, and WARR, L N. 1990. Upper Carboniferous foreland basin evolution in SW Britian. *Proceedings of the Ussher Society,* Vol. 7, 212–216.

HARWOOD, G M. 1976. The Staddon Grits — or Meadfoot Beds? *Proceedings of the Ussher Society,* Vol. 3, 333–338.

HAYWARD, A B, and GRAHAM, R H. 1989. Some geometrical characteristics of inversion. 17–40 *in* Inversion Tectonics.

COOPER, M A, and WILLIAMS, G D (editors). *Special Publication of the Geological Society of London*, No. 44.

HENDRIKS, E M L. 1937. Rock succession and structure in south Cornwall, a revision. With notes on the central European facies and Variscan folding there present. *Quarterly Journal of the Geological Society of London*, Vol. 93, 322–360.

HENNAH, R. 1816. Observations respecting the limestone of Plymouth. *Transactions of the Geological Society*, Series 1, Vol. 4, 410–412.

HIGGINS, A C, and AUSTIN, R L (editors). 1985. *A stratigraphical index of conodonts*. (Chichester: Ellis Horwood.)

HILL, J B, and MACALISTER, D A. 1906. The geology of Falmouth and Camborne and of the mining district of Camborne and Redruth. *Memoir of the Geological Survey of Great Britain*, Sheet 352 (England and Wales).

HILLIS, R R, and CHAPMAN, T J. 1992. Variscan structure and its influence on post-Carboniferous basin development, Western Approaches Basin, SW UK continental shelf. *Journal of the Geological Society of London*, Vol. 149, 413–417.

HINDE, G J, and FOX, H H. 1895. On a well marked horizon of radiolarian rocks in the Lower Culm Measures of Devon, Cornwall and west Somerset. *Quarterly Journal of the Geological Society of London*, Vol. 51, 609–668.

HINDE, G J, and FOX, H H. 1896. Supplementary notes on the radiolarian rocks in the Lower Culm Measures to the west of Dartmoor. *Transactions of the Devonshire Association*, Vol. 28, 774–789.

HOBSON, B. 1892. On the basalts and andesites of Devonshire, known as "Feldspathic Traps". *Quarterly Journal of the Geological Society of London*, Vol. 48, 496–507.

HOBSON, D M. 1976. The structure of the Dartmouth antiform. *Proceedings of the Ussher Society*, Vol. 3, 320–323.

HOBSON, D M. 1978. The Plymouth area. *Geologists' Association Guide*, No. 38.

HOLDER, M T, and LEVERIDGE, B E. 1986a. Correlation of the Rhenohercynian Variscides. *Journal of the Geological Society of London*, Vol. 143, 125–134.

HOLDER, M T, and LEVERIDGE, B E. 1986b. A model for the tectonic evolution of south Cornwall. *Journal of the Geological Society of London*, Vol. 143, 141–147.

HOLDER, M T, and LEVERIDGE, B E. 1994. A framework for the European Variscides. *Technical Report of the British Geological Survey*, WA/94/24.

HOLL, H B. 1868. On the older rocks of south Devon and east Cornwall. *Quarterly Journal of the Geological Society of London*, Vol. 24, 400–454.

HOLLOWAY, S, and CHADWICK, R A. 1986. The Sticklepath–Lustleigh Fault Zone: Tertiary sinistral reactivation of a Variscan dextral strike-slip fault. *Journal of the Geological Society of London*, Vol. 143, 447–452.

HOSKING, K F G. 1964. Permo-Carboniferous and later mineralisation of Cornwall and south-west Devon. 201–245 in *Present views on some aspects of the geology of Cornwall and Devon*. HOSKING, K F G, and SHRIMPTON, G H (editors). (Truro: Royal Geological Society of Cornwall.)

HOUSE, M R. 1963. Devonian ammonoid successions and facies in Devon and Cornwall. *Quarterly Journal of the Geological Society of London*, Vol. 70, 315–321.

HOUSE, M R, RICHARDSON, J B, CHALONER, W G, ALLEN W R L, HOLLAND, C H, and WESTOLL, T S. 1977. A correlation of Devonian rocks of the British Isles. *Special Report of the Geological Society of London*, No. 8.

HOUSE, M R, and SELWOOD, E B. 1964. Palaeozoic palaeontology in Devon and Cornwall. 45–86 in *Present views of some aspects of the geology of Cornwall and Devon*. HOSKING, K F G, and SHRIMPTON, G J (editors). (Truro: Royal Geological Society of Cornwall.)

HUMPHREYS, B, and KNOX, R W O'B. 1987. Investigations into the sedimentology of Devonian sediments of the Plymouth area. *Technical Report of the British Geological Survey, Stratigraphy Series*, SRG/87/14.

HUMPHREYS, B, and SMITH, S A. 1988. The sedimentology of a regressive–transgressive sequence: the Middle Devonian Meadfoot Beds, Staddon Grits and Jennycliff Slates, Plymouth Sound. *Technical Report of the British Geological Survey, Stratigraphy Series*, WH 88/256C.

HUMPHREYS, B, and SMITH, S A. 1989. The distribution and significance of apatite in Lower to Middle Devonian sediments east of Plymouth Sound. *Proceedings of the Ussher Society*, Vol. 7, 118–124.

HUNZIKER, J C, FREY, M, CLAUER, N, DALLMEYER, R D, FRIEDRICHSEN, H, FLEHMIG, W, HOCHSTRASSER, K, ROGGWILLER, P, and SCHWANDER, H. 1986. The evolution of illite to muscovite: Mineralogical and isotopic data from the Glarus Alps, Switzerland. *Contributions to Mineralogy and Petrology*, Vol. 92, 157–180.

INGLIS, J C. 1877. On the hydrogeology of the Plymouth district. *Journal of the Plymouth Institution*, Vol. 6, 105–121.

INSTITUTE OF GEOLOGICAL SCIENCES. 1977. Geological sheet 348 (Plymouth). 1: 50 000 Series. Institute of Geological Sciences.

INSTITUTE OF GEOLOGICAL SCIENCES. 1981. South Devon and Cornwall geochemical survey.

ISAAC, K P. 1983. The tectonic evolution of the eastern part of south-west England: Tectonothermal, denudation and weathering histories. Unpublished PhD thesis, University of Exeter.

ISAAC, K P, TURNER, P J, and STEWART, I J. 1982. The evolution of the Hercynides of central SW England. *Journal of the Geological Society of London*, Vol. 139, 521–531.

IVIMEY-COOK, H C. 1992. Devonian fish remains from 'The Long Stone', Cornwall. *Technical Report of the British Geological Survey, Stratigraphy Series*, WH/92/59R.

JACKSON, J, and MCKENZIE, D P. 1983. The geometrical evolution of normal fault systems. *Journal of Structural Geology*, Vol. 5, 471–482.

JAMES, H C L. 1981. Pleistocene sections at Gerrans Bay, south Cornwall. *Proceedings of the Ussher Society*, Vol. 5, 239–240.

JEAN, D M. 1984. Cave sites in the Plymouth Area. (Plymouth Caving Group, private publication.)

JONES, N S. 1992. Lacustrine and distal fan sediments and processes, Lower Devonian Dartmouth Group, SE Cornwall. *Technical Report of the British Geological Survey, Stratigraphy Series*, WH/92/111R.

JONES, N S. 1993. Sedimentology of Devonian and Carboniferous strata from the Plymouth area, south-west England. *Technical Report of the British Geological Survey, Stratigraphy Series*, WH/93/77R.

JONES, N S. 1995. Shallow marine sedimentation in the Lower Devonian Meadfoot Group (Bovisand Formation) from the Whitsand Bay area, with sedimentological notes on other Devonian strata examined within the area. *Technical Report of the British Geological Survey, Stratigraphy Series*, WH/95/210R.

JULIVERT, M. 1971. Decollement tectonics in the Hercynian cordillera of northwest Spain. *American Journal of Science*, Vol. 270, 1–29.

KIDSON, C. 1971.    The Quaternary history of the coasts of South-West England with special reference to the Bristol Channel.    1–22 in *Exeter essays in geography*. GREGORY, K J, and RAVENHILL, W L D (editors).    (Exeter: University of Exeter.)

KIDSON, C, and WOOD, T R. 1974.    The Pleistocene stratigraphy of Barnstaple Bay.    *Proceedings of the Geologists' Association*, Vol. 85, 223–237.

KISCH, H J. 1987.    Correlation between indicators of very-low-grade metamorphism.    227–300 in *Low temperature metamorphism*. FREY, M (editor).    (Glasgow: Blackie and Son.)

KLOMINSKY, J. 1994.    *Geological atlas of the Czech Republic: Stratigraphy.*    (Prague: Czech Geological Survey.)

KOSMAT, F. 1927.    Gliederung des variszischen gebirgsbaues. Abhandlung des Sachsichen Geologischen Landsamptes, Vol. 1. [German].    39 pp

KRAUSKOPF, K B. 1967.    *Introduction to geochemistry.*    (New York: McGraw-Hill.)

KSIAZKIEWICZ, M. 1954.    Graded and laminated bedding in the Carpathian flysch.    *Annales de la Societe geologique de Pologne*, Vol. 22, 399–449.

LAMING, D J C. 1970.    New Red Sandstone of Tor Bay, Petitor and Shaldon.    *Transactions of the Devonshire Association*, Vol. 101, 207–218.

LANE, A N. 1970.    Possible Tertiary deformation of Armorican structures in south-east Cornwall.    *Proceedings of the Ussher Society*, Vol. 2, 197–204.

LATTIMER, M. 1961.    Some prehistoric evidence in the Plymouth area.    *Transactions of the Devonshire Association*, Vol. 93, 288–303.

LEAKE, R C, BROWN, M J, SMITH, K, ROLLIN, K E, KIMBELL, G S, CAMERON, D G, ROBERTS, P D, and BEDDOE-STEPHENS, B W. 1985. Volcanogenic and exhalative mineralisation within Devonian rocks of the South Hams district of Devon.    *Mineral Reconnaissance Programme Report of the British Geological Survey*, No. 79.

LECKIE, D A. 1988.    Wave formed, coarse-grained ripples and their relationship to hummocky cross-stratification.    *Journal of Sedimentary Petrology*, Vol. 58, 607–622.

LECKIE, D A, and WALKER, R G. 1982.    Storm- and tide-dominated shorelines in Cretaceous Moosebar–Lower Gates interval — outcrop equivalents of deep basin gas trap in western Canada.    *Bulletin of the American Association of Petroleum Geologists*, Vol. 66, 138–157.

LE NEVE FOSTER, C. 1875.    Remarks on the lode at Wheal Mary Ann, Menheniot.    *Transactions of the Royal Geological Society of Cornwall*, Vol. 9, 152–157.

LEVERIDGE, B E, and HOLDER, M T. 1985.    Olistostromic breccias at the Mylor/Gramscatho boundary, south Cornwall. *Proceedings of the Ussher Society*, Vol. 6, 147–154.

LEVERIDGE, B E, HOLDER, M T, and GOODE, A J J. 1990.    Geology of the country around Falmouth.    *Memoir of the British Geological Survey*, Sheet 352 (England and Wales).

LISLE, R. 1992.    A new method of estimating regional stress orientations: application to focal mechanism data of recent British earthquakes.    *Geophysical Journal International*, Vol. 10, 276–282.

LITTLEJOHN, G S, and BRUCE, D A. 1979.    Long-term performance of high capacity rock anchors at Devonport. *Ground Engineering*, Vol. 10, 25–33.

MACALISTER, D A. 1921.    Total quantity of tin, copper and other minerals produced in Devonshire, particularly with regard to the amounts raised from each parish.    *Summary of Progress for 1920.* Geological Survey of Great Britain.    (London: HMSO.)

MACFADYEN, W A. 1970.    *Geological highlights of the West Country.* (London: Butterworth.)    296 pp.

MARTIN, H, and EDER, F W (editors). 1984.    *Intracontinental fold belts.*    (Berlin: Springer-Verlag.)

MARTINSSON, A. 1970.    Toponomy of trace fossils.    323–330 *in* Trace fossils. CRIMES, T P, and HARPER, J G (editors).    *Geological Journal Special Issue*, No. 3.

MATHESON, G D. 1983.    Rock stability assessment in preliminary investigations — graphical methods.    *Report of the Transport and Road Research Laboratory, Department of Transport*, LR 1039.

MATHESON, G D. 1988.    The collection and use of field discontinuity data in rock slope design.    *Quarterly Journal of Engineering Geology*, Vol. 22, 19–30.

MATTHEWS, S C. 1962.    A Middle Devonian conodont fauna from the Tamar Valley.    *Proceedings of the Ussher Society*, Vol. 1, 27–28.

MATTHEWS, S C. 1969.    A Lower Carboniferous conodont fauna from east Cornwall.    *Palaeontology*, Vol. 12, 262–275.

MATTHEWS, S C. 1970.    A new cephalopod fauna from the Lower Carboniferous of east Cornwall.    *Palaeontology*, Vol. 13, 112–131.

MATTHEWS, S C. 1977.    The Variscan fold belt in southwest England.    *Neues Jahrbuch fur Geologie und Palaeontologie*, Vol. 154, 94–127.

MATTHEWS, S C. 1984.    Northern margins of the Variscides in the North Atlantic region: comments on the tectonic context of the problem.    71–85 in Variscan tectonics of the North Atlantic Region. HUTTON, D W H, and SANDERSON, D J (editors). *Special Publication of the Geological Society of London*, No. 14.

MAYALL, M J. 1979.    Facies and sedimentology of part of the Middle Devonian limestones of Brixham, South Devon, England. *Proceedings of the Geologists' Association*, Vol. 90, 171–179.

MC BRIDE, E F. 1974.    Significance of color in red, green, purple, olive, brown, and gray beds of Difunta Group, Northeastern Mexico.    *Journal of Sedimentary Petrology*, Vol. 90, 171–179.

MCKENZIE, D, and BICKLE, M J. 1988.    The volume and composition of melt generated by extension of the lithosphere. *Journal of Petrology*, Vol. 29, 625–679.

MC NESTRY, A. 1993.    Devonian palynology of 8 samples from Torpoint and Tamar.    *Technical Report of the British Geological Survey, Stratigraphy Series*, WH/93/333R.

MC NESTRY, A. 1994a.    The palynology of 2 samples from the Devonian of 1:50k sheet 348.    *Technical Report of the British Geological Survey, Stratigraphy Series*, WH/94/145R.

MC NESTRY, A. 1994b.    A summary of the palynostratigraphy and general biostratigraphy of 1:50k sheets 348 and 349, Plymouth–Tamar area, Devon.    *Technical Report of the British Geological Survey, Stratigraphy Series*, WH/94/146R.

MERRIMAN, R J, EVANS, J A, and LEVERIDGE, B E. 2000.    Devonian and Carboniferous volcanic rocks associated with the passive margin sequences of SW England: some geochemical perspectives. *Geoscience in south-west England*, Vol. 10, 77–85.

MERRIMAN, R J, KEMP, S J, WARR, L N, and PRIOR, S R. 1996. Regional low-grade metamorphism in the Plymouth district, 1: 50 000 Sheet 348.    *Technical Report of the British Geological Survey*, WG/96/9.

MERRIMAN, R J, ROBERTS, B, PEACOR, D R, and HIRONS, S R. 1995. Strain-related differences in the crystal growth of white mica and chlorite: a TEM and XRD study of the development of metapelite microfabrics in the Southern Uplands thrust terrane, Scotland.    *Journal of Metamorphic Geology*, Vol. 13, 449–576.

MERSCHEDE, M. 1986.    A method of discriminating between different types of mid-ocean ridge basalts and continental tholeiites with the Nb–Zr–Y diagram. *Chemical Geology,* Vol. 56, 207–218.

MILLER, M F, and JOHNSON, K G. 1981. *Spirophyton* in alluvial-tidal facies of the Catskill deltaic complex: possible biological control of ichnofossil distribution. *Journal of Palaeontology,* Vol. 55, 1016–1027.

MITCHELL, G F. 1960. The Pleistocene history of the Irish Sea. *British Association for the Advancement of Science,* Vol. 17, 313–325.

MITCHELL, G F. 1973. The late Pliocene marine formation at St Erth, Cornwall. *Transactions of the Royal Society of London,* Vol. 266, 1–37.

MITCHELL, G F, and ORME, A R. 1967. The Pleistocene deposits of the Isles of Scilly. *Quarterly Journal of the Geological Society of London,* Vol. 123, 59–92.

MOLYNEUX, S G. 1990a. Palynological analysis of Devonian samples from 1:10 000 sheet SX 45 SE. *Technical Report of the British Geological Survey, Stratigraphy Series,* WH/90/74.

MOLYNEUX, S G. 1990b. Palynological investigation of samples from 1: 10 000 sheet SX 45 NW. *Technical Report of the British Geological Survey, Stratigraphy Series,* WH/90/268R.

MOLYNEUX, S G, and OWENS, B. 1990. Spores and acritarchs from samples of the Kate Brook Slates, Devon. *Technical Report of the British Geological Survey, Biostratigraphy Series,* PD/90/329.

MOORBATH, S. 1962. Lead isotope abundance studies on mineral occurrences in the British Isles and their geological significance. *Philosophical Transactions of the Royal Society, Section A,* Vol. 254, 295–360.

MUIR, I D, and RUST, B R. 1982. Sedimentology of a Lower Devonian coastal fan complex: the Snowblind Bay Formation of Cornwallis Island, Northwest Territories, Canada. 30, 245. *Bulletin of the Canadian Petroleum Geologists,* Vol. 263.

NEMEC, W, and STEEL, R J. 1984. Alluvial and coastal conglomerates: their significant features and some comments on gravelly mass flow deposits. 1–31 *in* Sedimentology of Gravels and Conglomerates. KOSTER, E H, and STEEL, R J (editors). *Canadian Society of Petroleum Geologists, Memoir,* No. 10.

OLDHAM, A D, OLDHAM, J E A, and SMART, J. 1978. *The limestones and caves of Devon.* (Private publication).

OLIVER, W A, and CHLUPAC, I. 1991. Defining the Devonian: 1979–1989. *Lethaia,* Vol. 24, 119–122.

OLSEN, H. 1987. Ancient ephemeral stream deposits: a terminal fan model for the Bunter Sandstone Formation (Lower Triassic) in the Tonder -3, -4 and -5 wells Denmark. 69–86 *in* Desert sediments: ancient and modern. FROSTICK, L E, and REID, I (editors). *Special Publication of the Geological Society of London,* No. 35.

ORCHARD, M J. 1975. Famennian conodonts and cavity infills in the Plymouth Limestones. *Proceedings of the Ussher Society,* Vol. 3, 49–54.

ORCHARD, M J. 1977. Plymouth–Tamar. 20–23 *in* A correlation of Devonian rocks of the British Isles. House et al. (editors). *Special Report of the Geological Society of London,* No. 8.

ORCHARD, M J. 1978. The conodont biostratigraphy of the Devonian Plymouth Limestone, south Devon. *Palaeontology,* Vol. 21, 907–955.

OSBORNE, R A L. 1983. Cainozoic stratigraphy at Wellington Caves, New South Wales. *Proceedings of the Linnean Society of New South Wales,* Vol. 107, 131–147.

OSBORNE, R A L. 1986. Cave and landscape chronology at Timor Caves, New South Wales. *Journal and Proceedings of the Royal Society of New South Wales,* Vol. 119, 55–75.

OWENS, B, MC NESTRY, A, and TURNER, N. 1993. Palynological report on ?Devonian–Carboniferous samples from various localities on 1:50 000 sheet 348. *Technical Report of the British Geological Survey, Stratigraphy Series,* WH/93/123R.

PEREZ-ALVAREZ, M. 1993. Mineralogical study of Sb mineralisation from Devon. *Technical Report of the British Geological Survey, Mineralogy and Petrology Series,* MPSR/93/11.

PINET, B, MONTADERT, L, MASCLE, A, CAZES, M, and BOIS, C. 1987. New insights on the structure and the formation of sedimentary basins from deep seismic profiling in Western Europe. 11–31 *in Petroleum geology of North West Europe.* BROOKS, J, and GLENNIE, K W (editors). (London: Graham and Trotman.)

PIPER, J D A. 1988. *Palaeomagnetic database.* (Milton Keynes: Open University Press.)

PLINT, A G. 1982. Eocene sedimentation and tectonics in the Hampshire Basin. *Journal of the Geological Society of London,* Vol. 139, 249–254.

PLINT, A G. 1983. Facies, environments and sedimentary cycles in the Middle Eocene Bracklesham Formation of the Hampshire Formation: evidence for global sea-level changes. *Sedimentology,* Vol. 30, 625–653.

POCKLEY, R P C. 1964. Four new uranium-lead ages from Cornwall. *Mineralogical Magazine,* Vol. 33, 1081–1092.

POLLARD, J E. 1976. 105–108 *in discussion of* A problematical trace fossil from the New Red Sandstone of South Devon. RIDGWAY, J M. 1974. *Proceedings of the Geologists' Association,* Vol. 85, 511–517.

POUND, C J. 1983. The sedimentology of the Lower–Middle Devonian Staddon Grits and Jennycliff Slates on the eastern side of Plymouth Sound, Devon. *Proceedings of the Ussher Society,* Vol. 5, 465–472.

RICE-BIRCHALL, B, and FLOYD, P A. 1988. Geochemical and source characteristics of the Tintagel Volcanic Formation. *Proceedings of the Ussher Society,* Vol. 7, 52–55.

RICHARDSON, J B, and McGREGOR, D C. 1986. Silurian and Devonian spore zones of the Old Red Sandstone continent and adjacent areas. *Bulletin of the Geological Survey of Canada,* No. 364.

RIDGWAY, J E. 1976. 108–109 *in discussion of* A problematical trace fossil from the New Red Sandstone of South Devon. RIDGWAY, J E. 1974. *Proceedings of the Geologists' Association,* Vol. 85, 511–517.

RILEY, N J. 1996. Examination of ammonoid samples collected by S Tunnicliff from 1:50 000 Plymouth Sheet 348. *Technical Report of the British Geological Survey, Stratigraphy Series,* WH/96/38R.

ROBERTS, J, CLAOUE-LONG, J, JONES, P J, and FOSTER, C B. 1995. SHRIMP zircon age control of Gondwana sequences in Late Carboniferous and Early Permian Australia. 145–174 *in* Non-biostratigraphical methods of dating and correlation. DUNAY, R E, and HAILWOOD, E A (editors). *Special Publication of the Geological Society of London,* No. 89.

ROLLIN, K E. 1988. A detailed gravity survey between Dartmoor and Bodmin Moor: the shape of the Cornubian granite ridge and a new Tertiary basin. *Proceedings of the Geologists' Association,* Vol. 99, 15–25.

ROXBURGH, I S. 1983. Notes on the hydrogeology of the Plymouth Limestone. *Proceedings of the Ussher Society,* Vol. 5, 479–481.

RUNDLE, C C. 1980. K-Ar ages for lamprophyre dykes from SW England. *Institute of Geological Sciences, Isotope Geology Unit Report,* No. 80/9 (unpublished).

RUNDLE, C C. 1981. K-Ar ages for micas from S W England. *Institute of Geological Sciences, Isotope Geology Unit Report,* No. 81/10 (unpublished).

SADLER, P M. 1973. An interpretation of new stratigraphic evidence from south Cornwall. *Proceedings of the Ussher Society,* Vol. 3, 52–55.

SCHMIDT, H. 1926. Schwellen- und Beckenfazies im ostrheinischen Palaeozoikum. *Zeitschrift der Deutschen Geologischen Gesellschraft,* Vol. 77, 226–234. [German]

SCRIVENER, R C, and BENNETT, M J. 1980. Ore genesis and controls of mineralisation in the Upper Palaeozoic rocks of North Devon. *Proceedings of the Ussher Society,* Vol. 5, 54–58.

SCRIVENER, R C, DARBYSHIRE, D P F, and SHEPHERD, T J. 1994. Timing and significance of crosscourse mineralization in SW England. *Journal of the Geological Society of London,* Vol. 151, 587–590.

SCRIVENER, R C, LEAKE, R C, LEVERIDGE, B E, and SHEPHERD, T J. 1989. Volcanic-exhalative mineralisation in the Variscan province of SW England. *Terra Abstracts,* Vol. 1, 125.

SCRIVENER, R C, LEVERIDGE, B E, GOODE, A J J, and DARBYSHIRE, D P F. 1995. The relationship of acid volcanism, elvans and granites in south-west England. (Abstract). *Proceedings of the Ussher Society,* Vol. 8, 457.

SCRIVENER, R C, SHEPHERD, T J, and GARRIOCH, N. 1986. Ore genesis at Wheal Pendarves and South Crofty Mine, Cornwall — a preliminary fluid inclusion study. *Proceedings of the Ussher Society,* Vol. 6, 412–416.

SCRUTTON, C T. 1977. Facies variation in the Devonian limestones of eastern south Devon. *Geological Magazine,* Vol. 114, 165–245.

SEAGO, R D, and CHAPMAN, T J. 1988. The confrontation of structural styles and the evolution of a foreland basin in central south-west England. *Journal of the Geological Society of London,* Vol. 145, 789–800.

SEILACHER, A. 1967. Bathymetry of trace fossils. *Marine Geology,* Vol. 5, 413–428.

SELWOOD, E B. 1982. The Devonian Rocks. 15–41 in *The geology of Devon.* DURRANCE, E M, and LAMING, D J C (editors). (Exeter: University of Exeter.)

SELWOOD, E B. 1990. A review of basin development in central south-west England. *Proceedings of the Ussher Society,* Vol. 7, 199–205.

SELWOOD, E B, EDWARDS, R A, CHESTER, J A, HAMBLIN, R J O, HENSON, M R, RIDDOLLS, B W, and WATERS, R A. 1984. Geology of the country around Newton Abbot. *Memoir of the British Geological Survey,* Sheet 339 (England and Wales).

SELWOOD, E B, THOMAS, J M, WILLIAMS B J, CLAYTON, R E, DURNING, B, SMITH, O, and WARR, L N. 1998. The geology of the country around Padstow and Camelford. *Memoir of the British Geological Survey,* Sheet 335/6 (England and Wales).

SHACKLETON, R M, RIES, A C, and COWARD, M P. 1982. An interpretation of the Variscan structures in SW England. *Journal of the Geological Society of London,* Vol. 139, 533–541.

SHELTON, A W. 1987. The structure of the Plymouth Limestone; a potential field investigation. Unpublished report for the British Geological Survey, University of Keele.

SHEPHERD, T J, MILLER, M F, SCRIVENER, R C, and DARBYSHIRE, D P F. 1985. Hydrothermal fluid evolution in relation to mineralisation in southwest England, with special reference to the Dartmoor-Bodmin area. 345–364 in *High heat production (HHP) granites, hydrothermal circulation and ore genesis.* (London: Institution of Mining and Metallurgy.)

SHEPHERD, T J, and SCRIVENER, R C. 1987. Role of basinal brines in the genesis of polymetallic vein deposits, Kit Hill–Gunnislake area, SW England. *Proceedings of the Ussher* Society, Vol. 6, 491–497.

SIMS, P, and TERNAN, L. 1988. Coastal erosion: protection and planning in relation to public policies — a case study from Downderry, south-east Cornwall. 231–244 in *Geomorphology in environmental planning.* HOOKE, J M (editor). (London: John Wiley and Sons.)

SMITH, I F, and ROYLES, C P. 1989. The digital aeromagnetic survey of the United Kingdom. *Technical Report of the British Geological Survey,* WK/89/5.

SMITH, N J P. 1993. The exploration of deep plays in the Variscan fold belt and its foreland. 667–675 in *Petroleum geology of northwest Europe.* Proceedings of the 4th Conference. PARKER, J R (editor). (London: The Geological Society.)

SMITH, S A, and HUMPHREYS, B. 1989. Lakes and alluvial sandflat-playas in the Dartmouth Group, south-west England. *Proceedings of the Ussher Society,* Vol. 7, 118–124.

SMITH, S A, and HUMPHREYS, B. 1991. Sedimentology and depositional setting of the Dartmouth Group, Bigbury Bay, south Devon. *Journal of the Geological Society of London,* Vol. 148, 235–244.

STEIGER, R H, and JAGER, E. 1977. Subcommission on geochronology: Convention on the use of decay constants in geo- and cosmochronology. *Earth and Planetary Science Letters,* Vol. 36, 359–362.

STEPHENS, F J. 1932. The ancient mining districts of Cornwall, Liskeard District. *Report of the Royal Cornwall Polytechnic Society,* Vol. 7, 159–178.

STEPHENS, N. 1966. Some Pleistocene deposits in North Devon. *Biuletyn Peryglacjalny Societas scientiarum Lodziensis,* Vol. 15, 103–114. [in English].

STEWART, I J. 1981a. The structure, stratigraphy and conodont biostratigraphy of the north eastern margin of Bodmin Moor and adjacent areas. Unpublished PhD thesis, University of Exeter.

STEWART, I J. 1981b. Late Devonian and Lower Carboniferous conodonts from north Cornwall and their stratigraphical significance. *Proceedings of the Ussher Society,* Vol. 5, 179–195.

STREEL, M, HIGGS, K, LOBOZIAK, S, RIEGEL, W, and STEEMANS, P. 1987. Spore stratigraphy and correlation with faunas and floras in the type marine Devonian of the Ardenne–Rhenish regions. *Review of Palaeobotany and Palynology,* Vol. 50, 211–229.

SYLVESTER, A G. 1988. Strike-slip faults. *Bulletin of the Geological Society of America,* Vol. 100, 1666–1703.

TAYLOR, P W. 1951. The Plymouth Limestone; and the Devonian tetracorals of the Plymouth Limestone. *Transactions of the Royal Geological Society of Cornwall,* Vol. 18, 146–214.

TEALL, J J H. 1888. *British petrography: with special reference to the igneous rocks.* (London: Dalau and Company.)

THOMAS, J M. 1988. Basin history of the Culm Trough of southwest England. 24–37 in *Sedimentation in a synorogenic basin complex.* BESLY, B M, and KELLING, G (editors). (Glasgow: Blackie.)

THOMPSON, A M. 1970. Geochemistry of color genesis in red-bed sequence, Juniata and Bald Eagle formations, Pennsylvania. *Journal of Sedimentary Petrology,* Vol. 40, 599–615.

TIDMARSH, W G. 1932. The Permian lavas of Devon. *Quarterly Journal of the Geological Society of London,* Vol. 88, 712–775.

TUNBRIDGE, I P. 1981a. Sandy high energy flood sedimentation — some criteria for recognition, with some examples from the Devonian of southwestern England. *Sedimentary Geology,* Vol. 28, 79–95.

TUNBRIDGE. 1981b. Old Red Sandstone sedimentation — an example from the Brownstones (highest lower Old Red Sandstone) of south-central Wales. *Geological Journal,* Vol. 16, 111–158.

TUNBRIDGE. 1984. Sandy model for a proximal stream and clay playa complex: the Middle Devonian Trentishoe Formation of north Devon. *Sedimentology*, Vol. 31, 697–716.

TURNER, N. 1993. A new approach to processing low-grade metamorphic argillaceous rocks of Devonian age from the Tamar Estuary and Plymouth districts. *Technical Report of the British Geological Survey, Stratigraphy Series*, WH/93/15R.

TURNER, P J. 1982. Aspects of the evolution of the Hercynides in central southwest England. Unpublished PhD thesis, University of Exeter.

USSHER, W A E. 1890. The Devonian rocks of south Devon. *Quarterly Journal of the Geological Society of London*, Vol. 46, 487–517.

USSHER, W A E. 1907. The geology of the country around Plymouth and Liskeard. *Memoir of the Geological Survey of Great Britain*, Sheet 348 (England and Wales).

USSHER, W A E. 1912. The geology of the country around Ivybridge and Modbury. *Memoir of the Geological Survey of Great Britain*, Sheet 349 (England and Wales).

VANDYKE, S, BERGERAT, F, and DUPUIS, C. 1991. Meso-Cenozoic faulting and inferred palaeostress in the Mons Basin (Belgium). *Tectonophysics*, Vol. 192, 261–271.

WANLESS, H R. 1979. Limestone response to stress: pressure solution and dolomitization. *Journal of Sedimentary Petrology*, Vol. 49, 437–462.

WARR, L N. 1993. Basin inversion and foreland basin development in the Rhenohercynian of south-west England. 197–224 in *The Rhenohercynian and Sub-Variscan fold belts*. GAYER, R A, GREILING, R O, and VOGEL, A K (editors). (Braunschweig/Wiesbaden: Vieweg Publishing.)

WARR, L N, PRIMMER, T J, and ROBINSON, D. 1991. Variscan very low-grade metamorphism in southwest England: a diastathermal and thrust related origin. *Journal of Metamorphic Geology*, Vol. 9, 751–764.

WATERS, R A. 1974. Palaeozoic successions and structures between Little Haldon and Dartmoor. Unpublished PhD thesis, University of Exeter.

WATSON, J, FOWLER, M B, PLANT, J A, and SIMPSON, P R. 1984. Variscan–Caledonian comparisons: late orogenic granites. *Proceedings of the Ussher Society*, Vol. 6, 2–12.

WEAVER, B L. 1991. Trace element evidence for the origin of ocean-island basalts. *Geology*, Vol. 19, 123–126.

WHITE, E I. 1956. Preliminary note on the range of Pteraspids in Western Europe. *Bulletin d' Institut royal des Sciences naturelles de Belgique*, Vol. 32, 1–10.

WHITE, R S. 1992. Crustal structure and magmatism of North Atlantic continental margins. *Journal of the Geological Society of London*, Vol. 149, 841–858.

WHITELEY, M J. 1981. The faunas of the Viverdon Down area, south-east Cornwall. *Proceedings of the Ussher Society*, Vol. 5, 186–193.

WHITELEY, M J. 1983. The geology of the St Mellion outlier, Cornwall, and its regional setting. Unpublished PhD thesis, University of Exeter.

WHITLEY, N. 1882. The evidence of glacial action in Cornwall and Devon. *Transactions of the Royal Geological Society of Cornwall*, Vol. 10, 132–141.

WILKINSON, I P. 1987a. Frasnian goniatites from Warren Point, near Plymouth. *Report of the British Geological Survey, Biostratigraphy Series*, PD/87/253.

WILKINSON, I P. 1987b. Devonian Entomozoacean Ostracoda and Cricoconarida in the neighbourhood around Plymouth. *Report of the British Geological Survey, Biostratigraphy Series*, PD/87/374.

WILKINSON, I P. 1987c. Devonian Ostracoda and Cricoconarida from the neighbourhood around Plymouth. *Report of the British Geological Survey, Biostratigraphy Series*, PD/87/428.

WILKINSON, I P. 1990a. Entomozoacean Ostracoda and Criconarida from the Upper Devonian purple and green slates of the Tamar estuary. *Report of the British Geological Survey, Stratigraphy Series*, WH/90/155R.

WILKINSON, I P. 1990b. A re-examination of Ussher's Entomozoacean Ostracoda and Cricocnarida from the Upper Devonian purple and green slates of the Tamar estuary. *Report of the British Geological Survey, Stratigraphy Series*, WH/90/156R.

WILKINSON, I P. 1993a. Calcareous faunas from Skinham Farm, near Carkeel. *Report of the British Geological Survey, Stratigraphy Series*, WH/93/118R.

WILKINSON, I P. 1993b. A summary of the distribution of Devonian Entomozoacean Ostracoda from the Tamar estuary. *Report of the British Geological Survey, Stratigraphy Series*, WH/93/157R.

WILKINSON, I P. 1993c. Microfaunas from a suite of samples from 1 inch Sheet 348. *Technical Report of the British Geological Survey, Stratigraphy Series*, WH/93/146R.

WILSON, J T. 1966. Did the Atlantic close and then re-open? *Nature, London*, Vol. 211, 676–681.

WINCHESTER, J A, and FLOYD, P A. 1977. Geochemical discrimination of different magma series and their differentiation products using immobile elements. *Chemical Geology*, Vol. 20, 325–343.

WORTH, R H. 1931. A new first chapter for the history of Plymouth. *Transactions of the Plymouth Institution*.

WORTH, R N. 1878. The palaeontology of Plymouth. *Journal of the Plymouth Institution*.

WORTH, R N. 1883. Some teeth from a Stonehouse bone cave. *Transactions of the Royal Geological Society of Cornwall*, Vol. 10, 165–168.

WORTH, R N. 1885. The raised beaches on Plymouth Hoe. *Transactions of the Royal Geological Society of Cornwall*, Vol. 10, 204–212.

WORTH, R N. 1888. Some detrital deposits associated with the Plymouth Limestone. *Transactions of the Royal Geological Society of Cornwall*, Vol. 11, 151–162.

WORTH, R N. 1891. 'Additional notes on Cornish Trias.' *Transactions of the Royal Geological Society of Cornwall*, Vol. 2, 341–345.

ZEUNER, F E. 1959. *The Pleistocene Period*. (London: Hutchinson.)

ZIEGLER, A M, and McKERROW, W S. 1975. Silurian marine red beds. *American Journal of Science*, Vol. 275, 31–56.

ZIEGLER, P A. 1982. *Geological atlas of western and central Europe*. (Amsterdam: Elsevier for Shell International Petroleum Maatschappij BV.)

ZIEGLER, W, and SANDBERG, C A. 1990. The Late Devonian standard conodont zonation. *Courier Forschungsinstituut Senckenberg*, Vol. 121, 1–115.

# FOSSIL INVENTORY

To satisfy the rules and recommendations of the international codes of botanical and zoological nomenclature, authors of cited species are listed below.

## Chapter 4    Devonian

### Dartmouth Group

Gastropods
*Bellerophon bisulcata* Salter

Vertebrates
*Althaspis leachi*
*Rhinopteraspis cornubica* McCoy, 1851
*Rhinopteraspis dunenisis* Roemer, 1855
*Rhinopteraspis leachi* White, 1938

### Meadfood Group    Bovisand Formation

Conodonts
*Caudicriodus curvicauda-celtibericus group* Carls and Gandl, 1969
*Icriodus bilatericrescens bilatericrescens* Zeigler, 1956
*Pandorinellina steinhornensis steinhornensis* Zeigler, 1956

### Meadfood Group    Staddon Formation

Brachiopod
*Spirifer hystericus* (Schlotheim 1820)

Miospore
*Emphanisporites schultzii* McGregor, 1973

Miscellanea
*Palaeophycus* Hall, 1847
*Rusophycus* Hall, 1852

### Tamar Group    Saltash Formation

#### Jennycliff division

Conodont
*Icriodus amabilis* Bultynck and Hollard, 1980

Miscellanea
*Spirophyton* Hall, 1863

#### Antony division

Brachiopods
*Longispina maillieux* Rigeaux
*Retichonetes armatus* Bouchard-Chantereaux in de Verneuil, 1845

Miospores
*Grandispora echinata* Hacquebard, 1957
*Rugospora flexuosa* (*Jushko*) Steel, 1974

#### Landulph division

Bivalve
*Posidonomya venusta* Münster, 1840

Conodonts
*Ancyrodella pristina* Khalymbadzha and Tschernyshera, 1970
*Palmatolepis glabra pectinata* Zeigler, 1962
*Palmatolepis marginifera marginifera* Helms, 1959
*Palmatolepis minuta minuta* Branson and Mehl, 1934
*Palmatolepis quadrantinodosalobata* Sannemann, 1955
*Polygnathus varca* (*varcus*) Stauffer, 1940
*Polygnathus nodocostatus* Branson and Mehl, 1934

Miospores
*Auroraspora asperalla* (Kedo) Van der Zwan, 1980
*Auroraspora hyalina* (Naumova) Streel, 1974
*Auroraspora macra* Sullivan, 1968
*Dictyotriletes submarginatus* Playford
*Diducites versabilis* (Kedo) Van Veen, 1981
*Emphanisporites schultzii* McGregor, 1973
*Grandispora cornuta* Higgs, 1975
*Raistrickia variabilis* Dolby and Neves, 1970
*Retispora lepidophyta* (Kedo) Playford, 1976
*Retusotriletes simplex* Naumova, 1953
*Rugospora flexuosa* (*Jushko*) Streel, 1974
*Vallatisporites pusillites* (Kedo) Dolby and Neves, 1970
*Vallatisporites verrucosus* Hacquebard, 1957

#### St Keyne division

Coral
*Pleurodictyum problematicum* Goldfuss, 1829

Trilobites
*Bradocryphaeus? cantarmoricus* Morzadec and Arbizu
*Greenops* (*Neometacanthus*) *perforatus* Morzadec

### Plymouth Limestone Formation

#### Faraday Road Member

Conodonts
*Icriodus corniger* Wittekindt, 1966
*Icriodus struvei* Weddidge, 1977

#### Prince Rock Member

Conodonts
*Icriodus expansus* Branson and Mehl, 1938
*Icriodus obliquimarginatus* Bischoff and Zeigler, 1957
*Icriodus regularicrescens* Bultynck, 1970

*Ozarkodina bidentata* Bischoff and Zeigler, 1957
*Ozarkodina brevis* Bischoff and Zeigler, 1957
*Polygnathus angustipennatus* Bischoff and Zeigler, 1957
*Polygnathus latus* Wittekindt, 1966
*Polygnathus linguiformis linguiformis* Hinde, 1879
*Polygnathus pseudofoliatus* Wittekindt, 1966
*Polygnathus trigonicus* Bischoff and Zeigler, 1957
*Polygnathus xylus ensensis* Ziegler, Klapper and Johnson, 1976

Corals
*Heliolites porosus* Goldfuss
*Macgeea bunthi* Taylor, 1951

### Plymouth Limestone Formation (undivided)

Conodonts
*Ancyrodella africana* Garcia-Lopez
*Icriodus alternatus* Branson and Mehl, 1934
*Icriodus latericrescens latericrescens* Branson and Mehl, 1938
*Ozarkodina brevis* Bischoff and Zeigler, 1957
*Palmatolepis delicatula* Ulrich and Bassler, 1926
*Palmatolepis disparaluca* Orr and Klapper, 1968
*Palmatolepis disparilis* Zeigler, Klapper and Johnson, 1976
*Palmatolepis superlobata* Branson and Mehl, 1934
*Palmatolepis triangularis* Sannemann, 1955
*Polygnathus ansatus* Ziegler and Klapper, 1976
*Polygnathus assymetricus unilabius* Huddle, 1934
*Polygnathus costatus costatus* Klapper, 1971
*Polygnathus cristatus* Hinde, 1879
*Polygnathus linguiformis linguiforms* Hinde 1879
*Polygnathus linguiformis klapperi* Clausen, Leuteritz and Zeigler, 1979
*Polygnathus linguiformis weddigei* Clansen Leuteritz and Zeigler, 1979
*Polygnathus timorensis* Klapper, Philip and Jackson, 1970
*Tortodus aff. variabilis* Bischoff and Zeigler, 1957

Corals
*Cylindrophyllum stonehousense* Taylor, 1951
*Macgeea varians*
*Mesophyllum thomasi* Taylor, 1951

Stromatoporoid
*Amphipora ramosa*

**Torpoint Formation**

Ammonoid (Goniatite)
*Manticoceras cordatum* Sandberger and
Sandberger

Ostracods
*Nehdentomis pseudorichterina* Matern, 1929
*Maternella dichotoma* Paeckelmann, 1913
*Maternella hemisphaerica* Richter, 1848
*Nehdentomis serratostriata* Sandberger,
1845
*Richterina costata* Richter, 1869
*Richterina striatula* Richter, 1848

## Chapter 5   Carboniferous

**St Mellion Formation**

Conodonts
*Elictognathus laceratus* Branson and Mehl,
1934
*Polygnathus communis communis* Branson
and Mehl, 1934
*Polygnathus triangulus* Voges, 1959
*Pseudopolygnathus radinus* (Cooper, 1939)

*Siphonodella cooperi* Haas, 1959
*Siphonodella duplicata* Branson and Mehl,
1934
*Siphonodella quadruplicata* Branson and
Mehl, 1934

Ammonoid (Goniatite)
*Ammonellipsites princeps* de Koninck
*Muensteroceras complanatum* de Koninck
*Nuculoceras naculum*
*Zephyroceras darwenense*

Miospores
*Lycospora pusilla* (Ibrahim) Somers, 1972

**Brendon Formation**

Conodonts
*Hindeodella undata* Branson and Mehl,
1941
*Polygnathus communis communis* Branson
and Mehl, 1934

**Newton Chert Formation**

Bivalve
*Posidonia becheri* Bronn

Conodonts
*Doliognathus lata* Branson and Mehl, 1941
*Gnathodus bilineatus* Roundy, 1926
*Gnathodus cuneiformis* Thomson and
Fellows, 1970
*Gnathodus girtyi* Haas, 1953
*Gnathodus pseudosemiglaber* Thomson and
Fellows, 1970
*Gnathodus semiglaber* Bischoff, 1957
*Gnathodus texanius* Roundy, 1926
*Hindeodella ibergensis* Bischoff, 1957
*Hindeodella segaformis* Bischoff, 1957
*Paragnathodus commutatus* Branson and
Mehl, 1941
*Pseudopolygnathus triangula pinnata*
Voges, 1959
*Scaliognathus anchoralis* Branson and
Mehl, 1941

Ammonoid (Goniatite)
*Neoglyphioceras spirale* Phillips

# INDEX

See also Contents (p.v) for principal headings and lithological units

## BRITISH GEOLOGICAL SURVEY

Keyworth, Nottingham NG12 5GG
0115 936 3100

Murchison House, West Mains Road, Edinburgh EH9 3LA
0131 667 1000

London Information Office, Natural History Museum
Earth Galleries, Exhibition Road, London SW7 2DE
020 7589 4090

Exeter Business Centre, Forde House, Park Five Business
Centre, Harrier Way, Sowton, Exeter, Devon EX2 7HU
01392 445271

The full range of Survey publications is available from the BGS
Sales Desks at Nottingham and Edinburgh; see contact details
below or shop online at www.bgs.co.uk

The London Information Office maintains a reference
collection of BGS publications including maps for
consultation.

The Survey publishes an annual catalogue of its maps and
other publications; this catalogue is available from any of the
BGS Sales Desks.

*The British Geological Survey carries out the geological survey of Great
Britain and Northern Ireland (the latter as an agency service for the
government of Northern Ireland), and of the surrounding continental
shelf, as well as its basic research projects. It also undertakes
programmes of British technical aid in geology in developing countries
as arranged by the Department for International Development and
other agencies.*

*The British Geological Survey is a component body of the Natural
Environment Research Council.*

## *the*StationeryOffice

Published by The Stationery Office and available from:

**The Stationery Office**
(mail, telephone and fax orders only)
PO Box 29, Norwich, NR3 1GN
Telephone orders/General enquiries 0870 600 5522
Fax orders 0870 600 5533

www.the-stationery-office.com

**The Stationery Office Bookshops**
123 Kingsway, London WC2B 6PQ
020 7242 6393  Fax 020 7242 6394
68–69 Bull Street, Birmingham B4 6AD
0121 236 9696  Fax 0121 236 9699
33 Wine Street, Bristol BS1 2BQ
0117 926 4306  Fax 0117 929 4515
9–21 Princess Street, Manchester M60 8AS
0161 834 7201  Fax 0161 833 0634
16 Arthur Street, Belfast BT1 4GD
028 9023 8451  Fax 028 9023 5401
The Stationery Office Oriel Bookshop
18–19 High Street, Cardiff CF1 2BZ
029 2039 5548  Fax 029 2038 4347
71 Lothian Road, Edinburgh EH3 9AZ
0870 606 5566  Fax 0870 606 5588

**The Stationery Office's Accredited Agents**
(see Yellow Pages)

*and through good booksellers*